What potential problems does biodiversity loss create for humankind? What basis is there for biologists' concern about what has been described as the sixth mass extinction on our planet? What costs does species extinction at both the local and global level impose on society? What can be done to prevent such loss? The Biodiversity Programme of the Royal Swedish Academy of Sciences' Beijer Institute brought together eminent economists and ecologists to consider these and other questions about the nature and significance of the problem of biodiversity loss. This volume reports key findings from that programme. In encouraging collaborative interdisciplinary work between the closely related disciplines of economics and ecology, programme participants hoped to shed new light on the concept of diversity, the implications of biological diversity for the functioning of ecosystems, the driving forces behind biodiversity loss, and the options for promoting biodiversity conservation.

The results of the programme are surprising. They indicate that the main costs of biodiversity loss may not be the loss of genetic material but the loss of ecosystem resilience and the insurance it provides against the uncertain environmental effects of economic and population growth. Because this is as much a local as a global problem, biodiversity conservation offers both local and global benefits. Since the causes of biodiversity loss lie in the incentives to local users, that is where reform must begin if the problem is to be tackled successfully.

Biodiversity loss

Biodiversity loss
Economic and ecological issues

Edited by

CHARLES PERRINGS
University of York

KARL-GÖRAN MÄLER
Beijer Institute

CARL FOLKE
Beijer Institute

C. S. HOLLING
University of Florida

BENGT-OWE JANSSON
University of Stockholm

CAMBRIDGE
UNIVERSITY PRESS

Published by the Press Syndicate of the University of Cambridge
The Pitt Building, Trumpington Street, Cambridge CB2 1RP
40 West 20th Street, New York, NY 10011–4211, USA
10 Stamford Road, Oakleigh, Melbourne 3166, Australia

First published 1995

Printed in the United States of America

Library of Congress Cataloging-in-Publication Data
Biodiversity loss: economic and ecological issues / edited by Charles
    Perrings.
        p. cm.
        "Outcome of the first research programme of the Royal Swedish
    Academy of Sciences' Beijer Institute"—Pref.
        Includes bibliographical references.
        ISBN 0-521-47178-8 (hc)
        1. Biological diversity conservation. 2. Biological diversity
    conservation – Economic aspects. I. Perrings, Charles. II. Beijer
    Institute.
    QH75.B5325 1995
    33.95'16 – dc20   94-27771
                                                                                      CIP

A catalog record for this book is available from the British Library

ISBN 0-521-47178-8 Hardback

Contents

v

Foreword

Resource economists typically view the natural environment through the lens of *population ecology*. Since the focus there is the dynamics of interacting populations of different species, it is customary to take the background environmental processes as exogenously given. The most well-known illustration of this viewpoint is the use of the logistic function for charting the time path of the biomass of a single species of fish enjoying a constant flow of food. Predator-prey models (e.g., that of Volterra) provide another class of examples: as do the May-MacArthur models[1] of competition among an arbitrary number of species.

One or more of the populations in question may have a value. The value may be utilitarian (e.g., as a source of food or as a keystone species), it may be aesthetic, or it may be intrinsic: Indeed, it may be all these things. In some cases the populations would be valued directly (we would then regard them as consumption goods), in others indirectly (in these cases they are capital goods). Depending on the context, the flow of value could be a function of the rate at which a population is harvested, or it could be a function of the population size; in many cases, it would be a function of both. For example, annual commercial profits from a fishery depend not only on the rate at which it is harvested, but also on the stock of the fishery, because unit harvesting costs are typically low when stocks are large and high when stocks are low. Rates at which populations are harvested under different institutional settings comprise the resource economist's object of inquiry. Of particular interest are optimal rates of harvest for a human community. Resource economics is thus a branch of what economists call the theory of capital.[2]

Environmental economists, on the other hand, often find it useful to

---

[1] See May (1972) and May and MacArthur (1972).

[2] The capital theoretic point of view is explored in Clark (1976) and Dasgupta (1982). For an application to fisheries, see Henderson and Tugwell (1979).

base their studies on *ecosystem ecology*. Here, the focus is on such objects as energy at different trophic levels and its rate of flow among them, the distribution and flows of biochemical substances in soils and bodies of water, and of gases and particulates in the atmosphere. The motivation here is to study the biotic and abiotic processes underlying the various functions that are performed by ecosystems. Holling's characterization of ecological processes as one of cycles of birth, growth, death, and renewal is a particularly illuminating example of this viewpoint (Holling, 1987, 1992). In such settings, the central concern of the economist, *qua* economist, is the valuation of the services that are provided by ecosystems. Economic studies of global warming, eutrophication of lakes, the management of rangelands, and the pollution of estuaries are typical examples of such endeavour (see e.g., Nordhaus, 1990; Costanza, 1991; Mäler et al., 1992; Walker, 1993).

Formally, population and ecosystem ecology differ by way of the state variables that are taken to characterize complex systems. In the former, the typical state variables are population sizes (or, alternatively, tonnage) of different species; in the latter, they are indices of the kinds of services mentioned above. It is often possible to summarize the latter in terms of the "quality," such as those for air, soil, or water. These ought all to be interpreted as summary statistics, reflecting as they do different forms of aggregation. Therein lies their virtue: They enable the analyst to study complex systems by means of a few strategically chosen variables.

The viewpoint just offered, that of distinguishing population and ecosystem ecology in terms of the state variables that summarize complex systems, permits us to integrate problems of resource management and those of environmental pollution and degradation (Dasgupta, 1982). In this way, insights from one field of study can be, and have been, used for gaining an understanding of others. The viewpoint also reminds us that ecological economics is concerned with the study of renewable natural resource systems when they are subject to human predation.

When described in such stark terms, both population and ecosystem ecology offer only partial viewpoints. For example, ecological processes are patently dependent on the composition of the biota, so implicit in ecosystem ecology is an account of population structures and the way they may be expected to alter over time through interactions both among themselves and with the abiotic processes at work. By the same token, population dynamics cannot be understood in the absence of knowledge of the relevant abiotic processes. Thus, the Clements-MacArthur proposal, if one may amalgamate the two perspectives, that species of organisms can be partitioned into those that are "*r*-strategists" (the pioneer, opportunistic species in recently disturbed habitats) and those that are

"*K*-strategists" (the conservative, or climax, species possessing strong competitive abilities), can be seen as a direct use of the logistic growth function mentioned earlier for the construction of a theory of ecological succession.[3] It follows that unified ecological models have both quantity and "quality" indices as state variables. It follows also that viewing the environmental resource base as a gigantic capital stock, as economists typically do, is a fruitful exercise.

The study of biodiversity is a testimony that population and ecosystem ecology are but partial lenses. Even ten years ago it was a popular belief that the utilitarian value of biodiversity was to be located in the potential future uses of genetic material for, say, pharmaceutical purposes. Preserving biodiversity was thus seen as a way of holding a diverse portfolio of assets with uncertain payoffs. The present collection of essays emphasizes an additional, possibly more important, value of biodiversity: as a source of resilience of ecosystems. Thus, field studies (e.g., Tilman and Downing, 1994) suggest that an ecosystem under stress (due, say, to the occurrence of violent events) manages to sustain much of its functions even when the composition of species changes. (Resilience should, therefore, be thought of as functional, as opposed to structural, stability of complex systems.) That is because there are species that are "waiting in the wings" to take over the functions of those that are denuded or destroyed. Now this is reminiscent of the assumption of "substitutability" among inputs in commodity production, an assumption that is often made in economic models of technological processes. But resilience presupposes that there *are* species waiting in the wings. So, to invoke the idea of substitutability among natural resources in commodity production in order to play down the utilitarian importance of biodiversity, as economists frequently do, is to engage in muddled thinking.

In one guise or the other, the essays in this splendid collection address these issues. The collection is the outcome of a project undertaken by the authors at the invitation of the Beijer International Institute of Ecological Economics in Stockholm. The programme involved both ecologists and economists. It was a privilege to have been allowed to attend the meetings at which these studies were planned and then presented. It was also a delight, because it was at these meetings that I saw how folk who spoke different languages gradually learned to understand one another and found that they could not only do business, but that the particular business of the day was both inspiring and pleasurable.

---

[3] If X is the population size of a species, its dynamics when growth can be characterized by the logistic function is: $dX/dt = r(X - X^2/K)$, where r and K are positive constants. For small values of X, the first term dominates, and the population experiences exponential growth, at the rate r. However, the long run stable population size is K.

REFERENCES

Clark, C. W. 1976. *Mathematical bioeconomics: The optimal management of renewable resources.* John Wiley, New York.

Costanza, R., ed. 1991. *Ecological economics: The science and management of Sustainability.* Columbia University Press, New York.

Dasgupta, P. 1982. *The control of resources.* Basil Blackwell, Oxford.

Henderson, J. V. and M. Tugwell. 1979. "Exploitation of the lobster fishery: Some empirical results," *Journal of Environmental Economics and Management,* 6, 287–96.

Holling, C. S. 1987. "Simplifying the complex: The paradigms of ecological function and structure," *European Journal of Operations Research,* 30, 139–46.

Holling, C. S. 1992. "Cross-scale morphology, geometry and dynamics of ecosystems," *Ecological Monographs,* 62, 447–502.

Mäler, K.-G. et al. 1992. "The baltic drainage basin programme," mimeo., Beijer International Institute of Ecological Economics, Stockholm.

May, R. M. 1972. "Will a large complex system be stable?" *Nature,* 238, 413–14.

May, R. M. and R. H. MacArthur. 1972. "Niche overlap as a function of environmental variability," *Proceedings of the National Academy of Sciences of the U.S.,* 69, 1109–13.

Nordhaus, W. 1990. "To slow or not to slow: The economics of the greenhouse effect," Discussion Paper, Department of Economics, Yale University.

Tilman, D. and J. A. Downing. 1994. "Biodiversity and stability in grasslands," *Nature,* 367(6461), January 27.

Walker, B. H. 1993. "Rangeland ecology: Understanding and managing change," *Ambio,* 22, 80–7.

Partha Dasgupta
Frank Ramsey Professor of Economics
University of Cambridge
and
Chairman
Beijer International Institute of
    Ecological Economics
Stockholm

Preface

This volume is an outcome of the first research programme of the Royal Swedish Academy of Science's Beijer Institute. The Institute was formed in 1991 with the primary goal of promoting interdisciplinary research between natural and social scientists in general, and ecologists and economists in particular, in an attempt to improve our understanding of the interdependency of economic and ecological systems. The programme invited a group of leading scholars in economics, ecology and related disciplines to consider the theoretical and policy issues associated with the biodiversity loss caused by direct depletion, by the destruction of habitat and by specialisation in agriculture, forestry and fisheries.

While it was motivated by the massive extinction of species, the Biodiversity Programme had a broader agenda than the problem of extinction. It addressed the driving forces behind all socially undesirable change in the composition of species and so went far beyond the protection of currently endangered species (much as their loss threatens the welfare of society). It included both the reasons for the overexploitation of environmental resources by millions of independent consumers and producers around the globe, and the scope for changing their behaviour so as to protect the ability of ecosystems to provide the ecological services on which humanity depends. This may be referred to as the problem of biodiversity conservation. It requires neither the preservation of all species, nor the maintenance of the environmental status quo. Instead it requires the development of the informational, institutional and economic conditions in which the private use of environmental resources may be made sustainable.

The programme focused on three sets of research issues. The first concerned the physical links between economic activities and change in the composition of species in an ecosystem, and the significance of that change in terms of its impact on the ecological services required for human consumption and production. The second concerned the reasons for the divergence between the private and social valuation of ecological services and, through this, the reasons for biodiversity loss. The third concerned the scope for minimising the costs in terms of biodiversity loss of damage already done to habitats, and the options open to reduce damage to other habitats in the future. All three concerns are reflected in this volume, although in their selection of papers the programme editors concentrated on those contributions that seemed to them to make the greatest strides either theoretically or methodologically.

The programme attracted the support of a group of social and natural scientists whose work in this and related areas is both well known and highly respected internationally. In all, 38 economists and ecologists contributed to the programme. Of these, 21 are represented in this volume. Other results of the programme have been or are in the process of being published elsewhere. In various ways their work has already changed our collective perception of where the problem in biodiversity lies, and what options we have in dealing with it. By shifting the emphasis from loss of genetic information to loss of ecosystem resilience, the programme has provided a powerful case for reappraising not just the costs of biodiversity loss, but the whole thrust of development based on specialisation of resource use. While research in this area has just begun, it is clear that it will have far reaching consequences for the management of biological resources – especially in those economies where high rates of population growth continue to increase demand for such resources. This is a startling achievement, and one that reflects the intellectual strength of the programme participants. But it also reflects the extraordinarily productive intellectual environment provided by the Royal Swedish Academy of Sciences, and the individuals appointed by the Academy to advise and direct the Beijer Institute. Whatever impact the programme has will be due as much to them as to programme participants.

Core funding for the programme was provided by the Beijer Foundation and the Swedish International Development Authority (SIDA). Additional sources of funding are acknowledged separately by the chapter authors. Joydeep Gupta and Tom van Rensburg provided technical editorial support.

Charles Perrings
York

Contributors

Edward B. Barbier
Senior Lecturer in Environmental
    Economics
University of York
Heslington, York YO1 5DD
Britain

Scott Barrett
Research Director at the Centre for
    Social and Economic Research on
    the Global Environment
    (CSERGE)
Associate Professor of Economics
London Business School
Sussex Place
Regents Park, London, NW1 4SA
Britain

I. J. Bateman
Lecturer in Environmental Economics
School of Environmental Sciences
University of East Anglia
Norwich, NR4 7TJ
Britain

Walter Boynton
Professor of Biology at the
    Chesapeake Biological Laboratory
University of Maryland
PO Box 38
Solomons, MD 20688
United States

Gardner Brown
Professor of Economics
University of Washington
Seattle, WA 98195
United States

Robert Costanza
President of the International Society
    for Ecological Economics
Director of the Maryland
    International Institute for
    Ecological Economics
Professor
University of Maryland
PO Box 39
Solomons, MD 20688
United States

Carl Folke
Deputy Director of the Beijer
    International Institute of Ecological
    Economics
Royal Swedish Academy of Sciences
PO Box 50005
S-104 05 Stockholm
Sweden

I. M Gren
Beijer International Institute of
    Ecological Economics
Royal Swedish Academy of Science
Box 50005
S-104 05 Stockholm
Sweden

C. S. Holling
Arthur R. Marshall Jr. Professor of
 Ecological Sciences
University of Florida
PO Box 118525
Gainesville, FL 32611
United States

Bengt-Owe Jansson
Director of the Stockholm Centre for
 Marine Research and Professor of
 Marine Ecology
Department of Systems Ecology
Stockholm University
S 106 91 Stockholm
Sweden

Michael Kemp
Professor of Biology
Horn Point Environmental Laboratory
University of Maryland
PO Box 775
Cambridge, MD 21613
United States

Karl-Göran Mäler
Director of the Beijer International
 Institute of Ecological Economics
Royal Swedish Academy of Sciences
PO Box 50005
S-104 05 Stockholm
Sweden

Charles Perrings
Director of the Beijer Institute
 Biodiversity Programme
Professor of Environmental
 Economics and Environmental
 Management
University of York
Heslington, York YO1 5DD
Britain

Michael Rauscher
Wissenschaftlicher Assistant
University of Kiel (and CEPR)
Institute of World Economics
D-24100 Kiel
Germany

Jonathan Roughgarden
Professor of Biological Sciences and
 of Geophysics
Stanford University
Stanford, CA 94305
United States

D. W. Schindler
Professor of Zoology
University of Alberta
Edmonton, T6C 2E9
Canada

Timothy Swanson
Lecturer and Research Director
Cambridge University
CSERGE, Faculty of Economics
Cambridge, CB3 9DD
Britain

R. K. Turner
Executive Director at the
Centre for Social and Economic
 Research in the Global
 Environment (CSERGE)
Professor of Economics
School of Environmental Sciences
University of East Anglia
Norwich, NR4 7TJ
Britain

Brian W. Walker
Director of the Division of Wildlife
 and Ecology
CSIRO
PO Box 84
Lyneham, Canberra, ACT. 2602
Australia

Martin L. Weitzman
Professor of Economics
Harvard University
Department of Economics
Cambridge, MA 02138
United States

Introduction: framing the problem of biodiversity loss

Charles Perrings, Karl-Göran Mäler, Carl Folke,
C. S. Holling, and Bengt-Owe Jansson

Identifying the problem in biodiversity loss

Of the two global environmental "problems" currently attracting almost obsessive popular interest, climate change and biodiversity loss, it is climate change that has dominated both the scientific and the policy agenda. The resources currently committed worldwide to both research on and mitigation of the causes of climate change all but eclipse the resources committed to biodiversity loss. If this fairly reflects the relative importance of the two problems it is difficult to avoid one of two conclusions. Either the social cost of biodiversity loss is trivial compared to the cost of climate change, or the return on resources invested in biodiversity conservation is much lower than the return on resources invested in arresting climate change. Either popular concern over the loss of biodiversity is misplaced, and the dire predictions of some of the world's best known biologists are not to be taken seriously, or the scope for reducing the social costs of biodiversity loss is small.

This volume considers just what is at issue in the problem of biodiversity loss. It does not engage in popular debate, nor does it enter the lists in the ranking of environmental problems. But it does ask questions that are central to the popular debate, and it does reach conclusions that cast doubt on the current commitment of resources to the biodiversity loss. The questions raised include those which have dominated the popular debate. What is the basis for biologists' concern with what has been described as the sixth mass extinction on our planet? What are the costs to society of both the local and global deletion of species? What can be done about it? But they go far beyond this. Some contributors to this volume ask what is the nature of biodiversity, and why is it ecologically significant? Others ask what is the incidence of the economic costs of biodiversity loss, and what does this mean for options open to decision-makers?

Some of the conclusions reached will come as a surprise to many scientists, and to almost all those – decision-makers included – who have engaged in the popular debate. In part this reflects the normal distance between the popular and scientific debate. In part it reflects the fact that scientific inquiry into the problem has barely begun. But most importantly, it reflects an unusual property of the work. The book reports the results of a programme of research in which ecologists and economists have combined in both setting the research agenda and addressing the questions it contains. Not only is this much less common an occurrence than one would expect, it also has a much greater impact on the outcome of the inquiry.

Our aim in this chapter is to sketch an outline of the ecological and economic problem in biodiversity loss and to provide an introduction to the treatment of the issue in the existing ecological and economic literature on biodiversity. This helps locate the studies that form the later chapters, and underlines why the questions they raise are important at this stage in our understanding of the problem. We do not pretend to offer a comprehensive treatment of the problem. Indeed, there is still massive ignorance and uncertainty about the extent and significance of change in the level of biodiversity. Extraordinarily little is known, for example, about even the existing diversity of species. Estimates of the total number of species on the planet range from five to one hundred million, of which less than one and a half million have been described, let alone analysed for their economically interesting properties (Miller et al. 1985). This volume offers little to reduce this sort of uncertainty. What it does is to identify the potential problems for humankind raised by the loss of biodiversity, and so to establish what questions it is worth posing in a world of finite resources. In so doing it reassesses the basis of the biodiversity debate, shifting our attention from the characteristics of particular organisms to the mix of organisms in ecosystems: from, for example, the rosy periwinkle to the composition of grasses in rangeland or of decomposer organisms in tropical moist forests. The traditional hunt for a cure for cancer in the specific properties of particular species gives way to a search for the role played by the mix of species and communities in maintaining the resilience of ecosystems from which we obtain valuable ecological services.

The following discussion reflects the general structure of the volume. The introduction and conclusions aside, the volume is divided into three parts. Part I addresses the conceptualisation of diversity, and the nature and role of biodiversity in ecological systems. Part II comprises a set of chapters in which economics and ecology have been integrated in the treatment of biodiversity loss in three ecosystem types, and Part III offers an economic perspective on the question of what can and needs to be done.

The ecological consequences of biodiversity loss

There are several levels at which it is possible to discuss biodiversity. Two have historically dominated the literature: genetic and species diversity. Recently, ecologists have begun to focus on a third level: diversity of ecological function. This reflects the growing recognition that species are economically interesting for two very different reasons. The first is the genetic properties which make distinct species of direct value in both human consumption and production. This is the basis for the traditional focus on both species and genetic diversity. In addition to this, however, species are interesting for the functions they perform in the generation of ecological services that are themselves of value to human society. That is, species also have indirect value through the ecological functions they perform. For some species, or for some groups of species, this value may be latent only. If there are a number of species capable of performing a certain function, not all of which perform that function under any given set of environmental conditions, those which are 'waiting in the wings' have what may be termed *insurance* or *option value*.

It follows that biodiversity may very well refer to rather different phenomena. Genetic, species and functional diversity are all distinct – albeit related – categories. Moreover, analysis of the problem of biodiversity loss in terms of one category does not necessarily give the same results as analysis in terms of another category, or in terms of two or more categories considered together. Nor is it clear from the existing literature that diversity necessarily means the same thing in each case. The characterisation of diversity functions is the problem addressed in Chapter 1 by Weitzman. The particular function used for illustration in that chapter recalls the phylogenetic trees of the biological literature on genetic diversity, but the choice of genetic difference as the criterion of distance is entirely arbitrary. The point Weitzman makes is that for any given criterion of distance between the different objects in some collection, it is possible to characterise a diversity function for that collection. It is also possible to construct a sequence of actions that minimises diversity loss by that criterion. This is of fundamental importance both for the conceptualisation of the biodiversity problem, and for the management of biodiversity loss.

The criterion of distance may be expected to vary depending on the problem under consideration, but it is obviously critical to get the right criterion. Genetic distance is one possible basis on which to construct a diversity function, but the implications for the management of biodiversity loss may be very different if one is considering other criteria of distance. Take, for example, the difference between domesticated and wild grasses of actual and potential use as staples. If the criterion of difference emphasises the properties of grasses that makes them of value in consumption, the distance between domesticated and wild grass species that are close substitutes in consumption would be slight. Disregarding the extra costs of maintaining the domestic grasses, this in a sense "authorises" the deletion of wild grasses, since little utility is lost in the process. Suppose, however, that the criterion of distance derives not from the properties of species that make them valuable in consumption, but from the properties of species that make them valuable in underwriting the provision of ecological services. If wild grasses perform the same ecological functions as domesticated grasses but under a different range of environmental conditions, and if those environmental conditions may occur with some probability greater than zero, then wild grasses have insurance value. They contribute towards the resilience of the system before shocks and stresses that alter those environmental conditions. Indeed, this is the primary motivation for biodiversity conservation in this volume. The criterion of distance implicit in most of the later chapters derives not from the genetic properties of species, but from the range of environmental conditions under which species can perform key ecological functions.

The theory that underpins the approach is described in Chapter 2, in which Holling et al. show how the structure and functioning of an ecosystem is sustained by the synergistic feedbacks between organisms and their environment. Species and their environments are connected in a web of interrelations that are fundamentally non-linear, with lags and discontinuities, thresholds and limits. The driving force in any system is the solar energy which flows through it, enabling the cyclic use of the materials and compounds required for the self-organisation and self-maintenance of the system. But it is the self-organising ability of the system, or more particularly the resilience of that self-organisation, which determines its capacity to respond to the stresses imposed by predation or pollution from external (including human) sources. The importance of biodiversity is argued to lie in its role in preserving ecosystem resilience, by underwriting the provision of key ecosystem functions under a range of environmental conditions.

Resilience, in the ecological literature, is understood in two rather different ways. One approach is associated with the ecosystem dynamics in the neighbourhood of (globally) stable equilibria. This approach concentrates on resistance to disturbance and speed of return to such equilibria. A second approach is associated with ecosystem dynamics where there exist multiple (locally) stable and unstable equilibria. This approach concentrates on the magnitude of the disturbance that can be absorbed before a system centred on one locally stable equilibrium passes (via an unstable manifold) into the basin of another. Holling et al. refer to the first as resilience of the first kind, and the second as resilience of the second kind. This volume is, in general, concerned with resilience of the second kind.

Ecosystem behaviour is described in terms of the sequential interaction between four system functions; exploitation (represented by those ecosystem processes that are responsible for rapid colonisation of disturbed ecosystems); conservation (as resource accumulation that builds and stores energy and material); creative destruction (where an abrupt change caused by external disturbance releases energy and material that have accumulated during the conservation phase); and reorganisation (where released materials are mobilised to become available for the next exploitive phase). The described pattern is discontinuous, and is characterised by the existence of multiple locally stable equilibria. Resilience is measured by the effectiveness of the last two system functions. It is crucial to the ability of the system to satisfy "predatory" demands for ecological services over time, and to cope with both sustained stress and shock, without losing local stability.

The arguments of Chapter 2 are not entirely uncontroversial. It is well known that the results reported on the link between the stability and complexity of ecosystems are contradictory (Begon, Harper and Townsend 1987; Orians and Kunin 1990). Holling et al. are able to report what appears to be an emerging synthesis, but this does not imply that there is any well defined general relationship between species diversity and ecosystem resilience. While it is clear that there are levels of biodiversity loss that cannot be sustained without inducing catastrophic change/fundamental reorganisation in all ecosystems (Ehrlich and Mooney 1983), it remains for future research to fix the boundaries to sustainable losses with greater precision than has been possible so far.

The critical point of much of the research being conducted along the lines described in Chapter 2 is that biodiversity matters primarily through its role in underpinning the performance of key ecological functions under varying environmental conditions. Just how system resilience depends on biodiversity varies from one system to another, but it is not the case that more species and more ecological structure always imply greater resilience. In Chapter 3, Costanza et al. review the characteristics and functions of biodiversity in coastal and estuarine ecosystems (with a special emphasis on the Chesapeake Bay in the United States). Except for coral reefs such systems tend to be low in species diversity. This is attributed by the authors to the unpredictability of the strong aperiodic physical forces of the estuarine environment which selects for generalist species. Whereas ecosystems in unpredictable terrestrial environments can build the structure needed to support higher levels of biodiversity, estuaries are so dominated by the physical forces of water flow that this is not possible. Estuaries are characterised by a small standing ecological structure and a high degree of organism mobility. Such ecosystems are highly resilient, in the sense of Chapter 2, but they do not show a high level of species diversity. The source of resilience is the ability of each species to support ecological functions under widely varying conditions. Using the Chesapeake Bay and its watershed as an exemplar they argue that what underpins the resilience of estuaries is the high level of functional diversity. The large and unpredictable physical forces in estuarine systems cause structural losses and keep high taxonomic diversity from developing, but the adaptability of the species that do exist ensures that the system can maintain its processes and productivity despite aperiodic and abrupt environmental change.

Integrating ecology and economics

We have already made the point that biodiversity satisfies human needs in two rather different ways. First, the individual organisms which collectively make up the biota have specific properties that make them of direct value in satisfying the consumption or production needs of human society. This is what lies behind the demand for particular species – their genes included. Second, the combination of organisms, and their role in sustaining biophysical cycles within the framework of a hierarchy of ecosystems, make them of indirect value in satisfying human needs for the services of those ecosystems. There is a loose correspondence between direct value and the value that biological diversity has to individual human users (its private value), and between indirect value and the value that biodiversity has to society (its social value). The existing literature tends to identify two different elements in private value. One element is argued to be the value of the resource in some definite use: its use value. The other element is argued to be a residual source of value, sometimes referred to as non-use value (Randall 1991) and sometimes as nonconsumptive value (Brown, 1990).

The private value of resources (having no nonconsumptive or nonuse value) is the value of the goods and services which the consumer is prepared to forgo by committing those resources to particular uses: the opportunity cost of those resources. In many cases, it is reasonably well approximated by the market price of those resources. The social value of resources, on the other hand, is very poorly understood. Since much of the social value of biodiversity is indirect, any estimate of such value depends on understanding how it contributes to the existence of general ecological services. Ecologists have begun the task of identifying the biophysical basis of the direct value of biodiversity (cf Ehrlich and Ehrlich 1992), but we have a good deal yet to learn about the biophysical basis of its indirect value. It is known that some subgroups of species – the keystone species – have stronger feeding interactions with each other than with the larger food web (Paine 1980). It is also known that some species – critical-link species – play a more "essential" role in ecosystem function than others, and that this is independent of their biomass, place in a food web, or possible role as a keystone species (Westman 1985). Indeed, such species are often found among decomposer microorganisms or litter invertebrates. Present ecological knowledge suggests that species play two major roles in ecosystems. First, they mediate energy and material flows, and so give ecosystems their functional properties. Second, they provide the system with the resilience to respond to events or surprises.

The net result is that while we can be sure that the indirect value of some species and ecosystems is greater than that of others, we have few reliable estimates of what the indirect value of the mix of species and ecosystems may be. At present, a clear distinction is made between the private nonuse value and indirect value of environmental resources. Nonuse value is conventionally defined as a residual: the difference between total value and use value. It has been argued at various times to include all of the following: bequest or scientific value (Krutilla 1967); the value conferred simply by the existence of a resource (Krutilla 1967; Pearce and Turner 1991); the value of the option to make use of the resource in the future (Weisbrod 1964); and the "quasi-option" value of the future information made available through the preservation of a resource (Arrow and Fisher 1974; Henry 1974a; Fisher and Hanemann 1986). Mäler (1991b) has argued that the distinction between the use and nonuse components of private value may be represented by a preference relation which exhibits certain separability properties. If, for some representation of human preferences in terms of a utility function, utility can be written as the sum of a function of the resource (say biodiversity) and a function which is not additive separable in the same resource and all other goods and services, then the nonuse value of biodiversity is simply the value associated with the first term, and the use value of biodiversity is the value associated with the second term. The problem is that while it is easy to demonstrate the existence of a component of value other than direct use value, it is not easy to explain this in a satisfactory way.

Although there is as yet no consensus on all of the elements in the value of biodiversity, it is clear that there is tremendous potential in a collaborative ecological-economic approach to the problem. The bases of the major elements – direct and indirect value – are understood in principle, even if they are not known in detail. In economics, recall that the valuation of environmental resources is an application of the theory of demand to the case where human welfare depends on the consumption not only of a basket of marketed goods, but also of some set of non-marketed environmental goods. In the present context, we seek the value of a change in the level of ecological services associated with a change in the level of biodiversity, where the ecological services are indirect and not marketed. Demand for these services may not be observed directly, but via the change in demand for market goods and services that accompanies change in the level of biodiversity.

The most theoretically satisfying of the available approaches to estimating demand for such unobserved services depends on the specification of a functional relationship between them and the marketed goods and services for which demand may be observed (cf Kolstad and Braden 1991). Three possible relations have been considered in the economic literature: the first, where marketed goods and ecological services are perfect substitutes; the second, where marketed goods and ecological services are perfect complements; and the third, where marketed goods and ecological services are weak complements. The perfect substitutes case implies that irrespective of the extent to which ecological services are lost, there always exists a level of defensive expenditures (a compensating variation in market income) that is capable of restoring utility (Freeman 1985; Mäler 1985). The perfect complements case implies the opposite, that there is no level of defensive expenditure which can compensate for the loss of ecological services. This case has not been explored in any depth. The third case has more than one variant. In the case discussed by Mäler (1974) it implies that the ecological service is weakly complementary with at least one marketed good. In the case discussed by Bockstael and McConnel (1983) it implies that the ecological service is weakly complementary with all final service flows generated by the household. Weak complementarity in both cases is defined to mean that the marginal utility of the ecological service is zero, when consumption of the weakly complementary market good or household services is zero. It implies the essentiality not of the ecological service, but of the market good. In both cases, too, it is possible to recover the change in welfare (the compensating variation in market income) associated with a change in the level of ecological services.

Where the ecological services in question are those associated with biodiversity loss it is immediately obvious that the case of perfect substitutes is inappropriate, but it is not immediately obvious that the other cases dealt with in the literature capture the relationship between the consumption of market goods and ecological services. To this point, very few attempts have been made to engage ecologists in the specification of the functional relationship between marketed goods and services and non-marketed ecological services. Indeed, this is one of the reasons why economists have paid so much attention to alternative but less theoretically satisfying approaches to the valuation of environmental resources (Smith, 1991).

There are still major problems in identifying the separate components of value. There is, for example, still no agreement on either terms or method. In Chapter 4, Turner et al. have addressed the problem of valuation for the case of wetlands in a way which attempts to break down direct and indirect value into a much finer set of categories. They characterise most use and nonuse values as "secondary" by contrast with the value of the stock of resources that is the ecosystem itself. This they term primary or "glue" value. Much of what we have referred to here as the indirect value of biodiversity relates to its contribution to the "glue" function of the ecosystem. There is some way to go before the methodology to identify this exists, but the first steps show that inclusion of the life-support function of ecosystems in the consumption and production functions used to describe economic activity has far reaching implications for the valuation of resources.

The remaining chapters in Part II are also examples of transdisciplinary collaborative work that, in one way or another, bear on the specification of production or consumption functions involving nonmarketed environmental resources, and draw attention to the importance of ecosystem functions in the behaviour of the economy. In Chapter 5, Roughgarden and Brown consider an economy based on the exploitation of a marine resource. The model they develop is somewhat different from models that have long dominated the fisheries literature, although the general analytical approach is similar. Interestingly, the conclusions reached in this chapter diverge very sharply from those in the standard fisheries models. Indeed, the authors conclude that from an ecological perspective, the perception of economic growth embodied in the basic neoclassical model is simply impossible. General ecosystem functions other than the provision of the stocks of interest are ignored, but the role of the ecosystem as nursery turns out to be critical to the results. That is, the optimal strategy turns out to be one in which the productive capacity of the broader ecosystem is maintained over time.

The model of rangeland utilisation developed by Perrings and Walker in Chapter 6 is somewhat less traditional in its analytical approach, though it too invokes the control framework that has long characterised work on the economics of natural resources. The problem addressed in the paper derives from the fact that the traditional Clementsian models of ecosystem dynamics turn out not to be very good predictors of ecosystem behaviour. Real ecosystems, as has already been observed, are nonlinear, discontinuous, and complex in their time-behaviour. There is no reason to believe that they will reconverge on a well-defined equilibrium (the climax state) following perturbation during the course of economic exploitation. Perrings and Walker consider the problem raised by the discontinuous biotic change that frequently occurs in managed semiarid rangelands. Discontinuous change is conceptualised as the transition from one locally stable rangeland state (to which corresponds one equilibrium mix of species) to another. The transition from a given state implies the loss of stability or resilience of that state. Using the state-and-transition model suggested by range ecologists as a framework, the paper discusses the appropriate decision-making process in this case, distinguishing between changes in states that can and cannot be known in advance.

Not all of these chapters explicitly address the value of system resilience. However, in contrast to traditional economic models of natural resource extraction, each recognises that what is ultimately being exploited is the productive capacity of the ecosystems generating such resources. Current modelling and analytical techniques are not well adapted to this shift in focus, and these chapters represent no more than preliminary forays into the field. However, they do point to one of the most distinctive characteristics of ecological economics, and one of the most important of the problems it addresses. The biomass from any given ecosystem may be thought of as a "flow" deriving from a "stock" that consists of the system itself. But this means that the "stock" is a changing phenomenon, evolving through adaptations to changes or fluctuations in the wider environment. The value of the "stock" depends on its productive capacity, which in turn depends on its adaptability. But this is a function of the resilience of the system, and so of the mix of the species it comprises. Indeed, the greater part of the indirect value of species may be said to lie in their contribution to the resilience of the system. Establishing the value of ecosystem resilience, and so the value of ecosystem components, is amongst the highest priorities on the ecological economics agenda.

One critical issue that waits on this is the determination of the maximum sustainable rate of discount. The ethics and rationality of discounting future costs and benefits are two of the most debated areas in economics (see Goodin 1982, for some of the points at issue). Moreover, given the significance of discounting for the allocation of environmental resources, it is not at all surprising that these debates have continued longer in environmental economics than elsewhere. It is easy to see where the moral qualms induced by discounting come from. The higher the rate of discount, the greater is the rate at which natural resources, including species, are optimally depleted. The higher the rate of discount, the less interesting are the costs of present activities visited on future generations. The higher the rate of discount, the more uncertainty is screened out of the information relevant to the decision-making process. The iron law of the discount rate – by which species with a growth rate less than the rate of discount will be optimally driven to extinction unless either the growth in the value of the species compensates for the difference, or their extraction is regulated – has brought more opprobrium on economists than almost any other proposition in economics. Nevertheless, the arguments of such as Randers and Meadows (1973) and Myrdal (1975) against discounting on the grounds that it induces individuals to inflict harm on future generations have fewer and fewer echoes in recent work, and those who still hold to the view (e.g., Daly and Cobb 1989) acknowledge that a low discount rate is not sufficient to prevent harm being inflicted on future generations.

It is now widely recognised that discount rates equal to the marginal productivity of the asset base may be ethically "neutral." The question remains, however, what does the marginal productivity of capital mean in circumstances where most of "capital stock" comprises zero-valued environmental resources – the ecosystems that underpin economic activity. Concern over the use of current real rates of interest to discount future environmental costs (quite rational if one is interested in the private valuation of those costs) derives from the fact that such interest rates bear no relation to any social rate of discount which might reasonably be authorised by the growth potential of the whole economy-environment system. But though it is intuitive that real rates of interest over five percent are "too high" from this perspective, it is not certain what social rate would be appropriate. It depends on the rate at which resources can be drawn from ecosystems without compromising ecosystem resilience, but this cannot be known without understanding the contribution of those resources to ecosystem resilience.

Biodiversity loss, economic theory, and economic policy

Most theoretical work on the economic concept of sustainability has drawn on a basic proposition in economics: that the maintenance of constant real consumption expenditure over time (maximum sustainable income in the sense of Hicks (1946), requires the maintenance of the value of the asset base (Mäler, 1991b; Solow 1986). The intuition behind this is that the level of consumption by any one generation may be repeated by the next generation only if the set of assets left to the next generation has at least equal productive potential (or value). Moreover, the level of consumption that is sustainable in this sense will be at a maximum, only if the set of assets left to the next generation has exactly the same productive potential (or value).

The main difficulty standing in the way of our understanding of the sustainability or otherwise of present patterns of economic activity is lack of information on the value of the asset base, given that this includes not just the produced capital whose market value is currently measured in the standard nation accounts, but also the nonmarketed natural capital whose value is implicitly assigned a zero weight in the accounts. A second difficulty, however, lies in the limitations of current economic models of the behaviour of the capital stock. Most of the existing models insulate the economic system from its environment and ignore the disequilibrating (evolutionary) tendencies within the system. The result has been a body of economic theory that has abstracted from what ecologists recognise as the most important characteristics of self-organising systems: the existence of environmental feedbacks, thresholds and discontinuities (nonconvexities). The dynamic interdependence of economic and environmental components of the system ensures that there will be feedbacks to any decision which leads to a change in the level of ecological services. Because of both the evolutionary nature of an ecosystem's responses to change in the level of economic stress and the existence of threshold effects, these feedbacks may be largely unpredictable except over ranges of stock sizes in which the ecosystem exhibits local stability (stays within the thresholds). Put another way, there are dynamic general equilibrium effects which (a) tend to be ignored in the process of aggregating values derived from partial observations of expenditure patterns given some change in the level of ecological services, and (b) are unpredictable except over the range of biodiversity in which ecosystems are stable.

In Chapter 7, Mäler explores the "requirements" of sustainable development in the context of an overlapping generations model. His conclusions emphasise the wide range of issues that economic and ecological science still has to address if it is to meet these requirements. These include the development of methods for predicting the future productivity of capital and the values of future damage to environmental resources including biodiversity, and for dealing with uncertainty about future damage. Such methodological improvements in the science are needed to identify the ethically neutral rate of time discount. Though this will still leave the problem of finding principles for risk sharing between generations. In addition, Mäler identifies a number of very practical policy and institutional requirements, of which the most significant is the development of institutions responsible for undertaking the necessary investments in order to compensate future generations for current resource degradation, and the development of incentives that ensure that it is in the private interest to conserve environmental resources.

Most loss of biodiversity is due to the independent decisions of the billions of individual users of environmental resources, and its underlying causes are to be found in the parameters within which those decisions have been made. These include the objectives that motivate decisions, the preferences that lie behind the demand for goods and services, the property rights that define individual endowments, the set of relative prices that determine the market opportunities associated with those endowments, and the cultural, religious, institutional and legal restrictions on individual behaviour that prescribe the range of admissible actions. Some organisms, such as the smallpox virus, have been driven to extinction "in the wild" because they have been perceived as a threat to human welfare. Some, such as the Moa, have been driven to extinction because of their desirable consumption properties. For most, however, extinction has been the incidental and usually unanticipated consequence of an economic activity that has destroyed their habitat – as is currently the case with the thousands of species being driven to extinction annually due to the destruction of tropical rainforests and coral reefs (Myers 1988).

The problem is that while the individual decisions that have led to most extinctions have been privately rational given the information available to the decision-maker, it is most doubtful whether they represent the best outcome for society. One cause of this is the noncorrespondence between the structure of property rights and real flows within the economy-environment system. This is the classic cause of external effects. Since market transactions depend upon a prior allocation of property rights, an incomplete allocation of rights implies that there will be effects of economic activities that are not mediated by the market. A second cause is the distortion of market prices as a result of government policy or strategic market behaviour. Rational individuals faced with market prices that do not reflect the social opportunity cost of the action – whether the cause is missing markets or misguided government intervention – will necessarily make decisions that are socially suboptimal. A separate source of difficulty is a distribution of income and assets that deepens the wedge between the individual private and social valuation of ecological services of a large part of humanity. Both the information used in private decisions and the rate at which the future consequences of those decisions are discounted are sensitive to the level of market income. Information, as Dasgupta (1990) has remarked, is not costless. The poor are able to command less information than the rich. In addition, there is a strong relationship between income and the rate at which people discount the future. Because what matters is consumption today, people in poverty will tend to discount the future costs of resource use at a very high rate (Perrings 1989).

If we take the first of these sources of divergence between private and social cost, environmental external effects are a consequence of two things: the dependence of all those affected by the transaction on a common environment, and the absence of some forum (market) in which to negotiate the value of those effects. Dependence on a common environment does not, of course, imply that all parties are affected equally by actions which generate externalities. Following Dasgupta (1991), it may be helpful to distinguish between two types of externality. Reciprocal externalities are said to be those in which all parties having rights of access to a resource are able to impose costs on each other. The short run external environmental costs or benefits of resource use are the same irrespective of who is responsible (this is the classical problem of the commons). Unidirectional externalities are said to be those in which the hydrological or other cycles of the common environment ensure that the short run external environmental costs or benefits of resource use are "one way" (for example, deforestation by the users of an upper watershed inflicts damage on the users of the lower watershed). Since all users are part of the same set of biogeochemical cycles the term "unidirectional" should not be taken too literally. The point is that the external costs and benefits of resource use in the two cases will be asymmetrical, and the "solutions" to each type of externality are rather different.

Externalities, in general, are evidence of the incompleteness of markets, and so of the incompleteness of the structure of property rights embedded in the institutional framework of society. They will tend to be prevalent wherever the structure of property rights is such that individual users are authorised to ignore the costs they impose on others. The issue we wish to raise here concerns the nature of property rights in species which give rise to unidirectional externalities involving species loss, and the incentive effects of different property rights. It is clear that there is a very wide range of rights currently conferred by law or custom on the users of biotic resources, and that these rights vary widely both from one country to another, and from one species to another. For example, rights with respect to migratory or otherwise mobile organisms differ from those with respect to stationary organisms. Most important of all, however, is the fact that many ecological services are neither rival nor exclusive in consumption. Consumption of such services by one user or group of users does not diminish their availability to others. Nor does consumption by one user or group of users preclude their consumption by others. They are in the nature of public goods.

It is often the case that populations within a single species may be subject to exclusive rights by an individual economic agent. Livestock farmers, for example, typically enjoy exclusive rights in their own herds. It is less often the case that individuals enjoy exclusive rights in the ecological services provided by the wider mix of species and populations within an ecosystem – the biodiversity of that system. Ownership of a herd generally imposes costs or confers benefits on others through the role of the herd in determining the broader flow of ecological services provided by the system. Indeed, the greater the size and connectedness of the ecosystems to which the directly exploited species belongs, the less is the potential for individuals to enjoy exclusive rights in the ecological services generated by the mix of species and populations in the system. Biodiversity is, in all the most important cases, a public good. There is a systematic bias against private investment in diversity, and in favour of investment in the specific populations whose benefits can be captured.

The remaining chapters in Part III deal with the implications of this at the international level. In Chapter 8, Swanson argues that the cause of inefficient diversity depletion is the decentralised process of land use conversions, by which he means that land use conversion decisions are undertaken on a national rather than an international basis. National decision-makers fail to take into consideration the presence of externalities from the depletion of the global stocks of diversity, and so convert land at an excessive rate. Swanson argues that an optimal policy for global biodiversity requires international intervention in the conversion decisions of individual states in order to halt the global conversion process at a stage prior to the unregulated result. He recommends the creation and maintenance of a "global premium" on investments in diversity. The premia should, he argues, be funded by the international community and directed towards the national states. More particularly, they should take one of two generic forms: international subsidy agreements or market regulation.

In Chapter 9, Barbier and Rauscher address the question of which of these is most likely to be effective in the case of tropical deforestation. Current rates of tropical deforestation are excessive for precisely the reasons discussed by Swanson. The social cost of the consequential loss of biodiversity is ignored in the decisions of those responsible for felling timber. Barbier and Rauscher consider two options for intervention: the use of taxes and tariffs to shift the terms of the international timber trade against the export of timber, and the use of international transfers. Their results suggest that trade interventions are, in the long run, a second-best policy option, and that transfers are likely to be more effective.

This accords with the widespread perception that the environment is best served by trade liberalisation and aid promotion (see, for example, Agenda 21: UNCED, 1993). We do not comment on this here, save to note that aid, in this context, is the "price" of international cooperation. Hence aid promotion is an important component of the Biodiversity Convention. It is intuitive that the international problem has at least elements of the prisoner's dilemma. Perceived as a noncooperative game, the role of the Convention is to secure the cooperative outcome. That is, if the structure of payoffs to habitat conversion in each of a number of non-cooperating nation states is such that the equilibrium outcome is worse than might be obtained through cooperation, there may exist a variation in the structure of payoffs that will induce cooperation.

In Chapter 10 Barrett considers the problem in these terms focusing on the conditions in which agreements such as the Biodiversity Convention might be self-enforcing with transfers. He concludes that agreement can sustain the full cooperative outcome only where the global net benefits of cooperative behaviour are "slightly larger" than the global net benefits of noncooperative behaviour. If one believes that the difference between the cooperative and noncooperative outcomes is very significant, then Barrett's conclusions are rather depressing. The wider the gap between the appropriable local and nonappropriable global benefits of biodiversity conservation, the less the likelihood that the cooperative outcome will be self-enforcing.

Biodiversity and resilience revisited

The critical questions from an economic policy perspective concern the distribution of the benefits of biodiversity conservation as between individual users of biological resources, nation states, regional groupings and the international community. It is the fact that the private and social costs of biodiversity depletion differ that creates the problem of excessive biodiversity loss in the first instance. But, as Barrett shows, it is the distribution of the social costs of depletion that determines how effectively the problem may be addressed. The link that is now being emphasised between functional diversity and ecological resilience is significant precisely because of its implications for the distribution of the social costs of biodiversity depletion. If the main implications of biodiversity depletion lie in the genetic information losses associated with species or population deletion, then it follows that the main social cost of biodiversity loss is indeed global and Barrett's results are discouraging. However, if the main implications of biodiversity depletion lie in the loss of ecosystem resilience, then the main social costs of biodiversity loss will be ecosystem-specific. That is, to a large extent they will manifest themselves in the same physical system as that in which the losses were incurred.

It is hard to overstate the significance of this observation. If it is understood that a large part of the social costs of biodiversity are local, not only is the problem easier to address within the nation state, but the relative value of the transfers needed to induce international cooperation is reduced. Species deletion at private hands may still involve both information and insurance costs that are incurred by the global community, but if the main external costs of private actions leading to loss of functional diversity accrue to other users of the same ecosystem, the optimal policy response is a local one. We return to this issue in the concluding chapter of this volume. It is, however, worth flagging the point at the outset since it is this, more than anything else, that will seem surprising to those whose attitudes to the biodiversity issue have been shaped by the popular debates on the extinction of species.

PART I

CONCEPTUALISING DIVERSITY AND
ECOSYSTEM FUNCTIONS

CHAPTER 1

Diversity functions

Martin L. Weitzman

1.1 Introduction

"Loss of diversity" is a much lamented condition nowadays. One sees such a phrase applied loosely in a variety of contexts, including the realms of biological species, landmark buildings, historic sites, languages, artifacts, habitats, even ways of life. Often there is an implicit injunction to preserve diversity because it represents a higher value than other things, which by comparison are "only money". Yet the laws of economics apply to diversity also. We cannot preserve everything. There are no free lunches for diversity. Given our limited resources, preservation of diversity in one context can only be accomplished at some real opportunity cost in terms of well-being forgone in other spheres of life, including, possibly, a loss of diversity somewhere else in the system.

Actual implementation of any injunction to "preserve diversity" is hampered by the lack of an operational framework or objective function. We need a more-or-less consistent and usable measure of the value of diversity that can tell us how to trade off one form of diversity against another.

It would be naive to expect that resolution of real-world conservation choices will reduce to some mechanical application of diversity functions. Yet, I would argue, it is still useful to think in terms of a model that might serve as a paradigm for guiding and informing conservation decisions, even if the model must be at a high level of abstraction. When diversity cannot be defined even under ideal circumstances, the concept itself is suspicious. For this reason alone, it behooves us to specify a diversity function at least for some "ideal" case.

If a value of diversity function can be meaningfully postulated, then it can, at least in principle, be made commensurate with other benefits and costs, and the general form of the resource allocation problem is in principle well defined. There are presumably some limits on the feasible actions that can be taken, represented by budget constraints or other limitations. Each feasible action induces a probability distribution for what survives, and for how long (there is, perhaps, a significant amount of correlation involved). The optimal conservation policy may be defined as the feasible action that yields the highest present discounted expected value of diversity (plus whatever other net benefits are attributed to various components). This is in the form of a classical constrained optimisation problem. Since the constraint set is in principle well defined, the major unresolved conceptual issue involves defining a meaningful value-of-diversity objective function. The remainder of the paper concentrates on this critical aspect of the problem.

1.2   The nature of the problem

There is an immediate issue of defining the proper unit of analysis for the collection whose diversity is to be determined. It is not transparently clear in all conservation settings at what level the diversity problem should be attacked. In principle, diversity could be measured at the individual level, the species level, the community level, the ecosystem level, or even some other levels. (The Nature Conservancy usually takes its "mapping units" or "elements" to be species or communities.) I do not have a good resolution to the problem of which level is the most appropriate for performing diversity analysis. In principle, any level might be chosen so long as the methodology is consistently followed at that level.

The abstract form of the general problem can now be stated. The "elements" are basic units that it is desired be preserved in the name of diversity. (There could also be some direct net benefits from some elements.) There is some notion of the joint probabilities of extinction of the various elements if no action is taken. Next, there are preservation "actions" that can influence the various probabilities of extinction at some cost. For example, projects might consist of buying up and preserving various specific sites. Some "diversity function" evaluates the diversity of any given deterministic collection of elements. Conceptually, the diversity function, on which this paper concentrates, is the most difficult part of the problem. If a diversity function is well defined, and probabilities of extinction are known, an expected diversity function can be defined. An expected diversity function is basically the sum of the deterministic diversity function of various collections of species weighted by the existence probabilities of the various collections. The basic aim might be taken to maximise expected present discounted diversity (plus any net direct values of the elements), subject to conservation budget constraints. The set-up is analogous to a capital budgeting problem, except that the objective contains expected diversity.

For convenience and consistency, in what follows I will largely employ biological metaphors. However, the mathematical essence of the problem applies to a broader setting and is perhaps more appropriately understood at a higher level of abstraction. The basic underlying unit will be called a "species." A "species" could stand for a genuine species in the traditional biological sense of being a reproductively isolated group having a history of strict genetic divergence from other groups; or it could stand for an individual, a subspecies, a specimen, an object, a community, or almost anything else – depending on the context.

Actually, one of the most useful interpretations is that a "species" corresponds to a "library." Conceptualising the basic problem in terms of preservation priorities among "libraries" is useful for at least two reasons. First of all, a library, at least in the abstract, tends to be a more or less neutral object that does not conjure up such strong emotional images as some other metaphors. Second, and more significantly, concentrating on libraries and the books they contain can help us to focus more sharply on the issue of what should be meant by a diversity function.

In what follows, then, the word "species" may be interpreted as some generalisation of the word "library." The reader who wants a specific image may find it useful to think of a species as standing for a library.

Suppose, then, there is some set S containing n member species (or n libraries). The basic question is how to measure the diversity of S. The appropriate diversity function will be denoted in this paper $V(S)$.

It is important to realise that there is unlikely to be a universally best definition of diversity. This is just common sense reasoning by analogy. There is not in statistics a universally best definition of central tendency, or of dispersion. Nor is there in economics a universally best definition of income inequality, or of welfare, or of industrial concentration. The appropriate definition depends upon the assumptions behind the specific intended application. However, just as in statistics or economics the field tends in practice to narrow down to only a few good candidates, so too I will try to argue, it is not so easy to find many good candidates for a diversity function. I will try to explain and justify my own formulation of a diversity function. Although I have not yet come across a formulation that satisfies me as much as the one I present here, the field is young and it is not to be excluded that some fresh approach might yield new insights.

What should one mean by a "good" definition of a diversity function? I think, as with the quantification of any concept, there are two general criteria.

First of all, the definition should be *a priori* sensible in that it embodies an intuitively plausible formulation that does not immediately admit of seriously damaging counterexamples to the basic underlying idea. Obviously, this criterion contains a subjective element. Second, and perhaps more critically, there should be some special case, hopefully a *sensible* special case, forming the central paradigm, for which the particular formulation is exactly the right answer to a rigorously well-posed problem. The definition in the general case then becomes seen as an appropriate abstraction of the basic concept to a situation where the problem is less rigorously stated than it is in the central paradigm.

The approach described above is consistent with standard statistical methodology. Basically, there is a rigorous model, which works exactly for an idealised situation that is not excessively bizarre even though it may not precisely characterise the real world. Additionally, the model itself seems sensible on heuristic grounds for the general case. I cannot here test the robustness of the model itself, because that would require a more general meta-model with a more general meta-definition of diversity, which I do not have. The most that can be said at this stage is that my definition of diversity works exactly for a nonbizarre central case, it has some nice properties and makes heuristic sense in the general case, and it might be hoped to have some robustness properties if one knew how to formulate them properly.

I want to start with the rigorous model of a nonbizarre special case. That is, I want to lay out a particular model of an idealised situation where it is really fairly clear what we should mean by a diversity function. The particular model is what I will call the "bead model" of an evolutionary branching process.

1.3   The bead model of an evolutionary branching process

Consider a treelike branching process such as depicted in Figure 1.1. The "species" 1 through 6 are depicted as twig tips at the end of the tree. Species evolve by descent with modification via an evolutionary branching process, which is described as follows.

Any species consists of the same very large number M of tiny beads strung together on a string. If the primary interpretation of a "species" is a "library," then the "beads" stand for "books." A species is essentially identified with its string of beads, just as a library is identified with its collection of books. The beads are accumulated over time by being drawn from an infinitely large sample pot of different beads. At each unit of time, for each species existing at that time, exactly one bead is independently drawn from the infinitely large sample of different beads and attached to the head of the string. Simultaneously, exactly one bead is dropped from

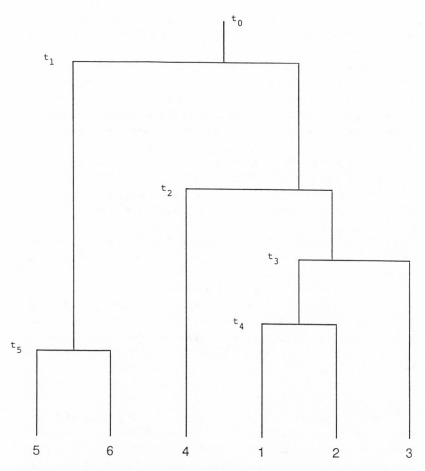

Figure 1.1. The maximum-likelihood tree representation.

the tail of the string. (Other descriptions are possible, but this one is the simplest.)

At time t_0, just one prototype ancestor species exists. From time t_0 to time t_1 exactly $t_1 - t_0$ different new beads are accumulated at the head of the string, while the same number of old beads are discarded from the tail.

Then at time t_1, a bifurcation into two ancestor species occurs. One of these is the common ancestor of $\{5, 6\}$. The other is the common ancestor of $\{1, 2, 3, 4\}$. Each of these two ancestor species, which are thought of as separate, now begins *independently* to accumulate different new beads at the head of the string, one per unit time, while simultaneously discarding one bead per unit time from the tail of the string. The next bifurcation occurs at time t_2. At that time, the ancestor species of $\{5, 6\}$ and of $\{1, 2, 3, 4\}$ differ by exactly $t_2 - t_1$ beads, which they have independently acquired during the time duration $t_2 - t_1$.

At time t_2, the common ancestor of $\{1, 2, 3, 4\}$ bifurcates into two species. One of these is the common ancestor of $\{4\}$. The other species is the common ancestor of $\{1, 2, 3\}$. From time t_2 to t_3, each of the three ancestor species $\{5, 6\}$, $\{4\}$, and $\{1, 2, 3\}$ is *independently* accumulating different new beads at the head of their strings, one per unit time, while simultaneously discarding one bead per unit time from the tails of their strings. This phase ends at time t_3, when ancestor species $\{1, 2, 3\}$ bifurcates into ancestor species $\{1\}$ and ancestor species $\{2, 3\}$. At that time t_3, ancestor species $\{1, 2, 3\}$ differs from ancestor species $\{4\}$ by $t_3 - t_2$ beads, while ancestor species $\{1, 2, 3\}$ and ancestor species $\{4\}$ both differ from ancestor species $\{5, 6\}$ by $t_3 - t_1$ beads.

The evolutionary branching process described above ends at the present time with the six currently existing species shown in Figure 1. 1. The last bifurcation that occurred was at time t_5, when ancestor species $\{5, 6\}$ split into ancestor species $\{5\}$ and ancestor species $\{6\}$.

The model of an evolutionary branching process described above is of course an idealisation. Actual evolution differs in many important ways. Nevertheless, as an abstraction, the bead model captures the essential idea of descent with independent modification along reproductively isolated lineages. Supposing for the sake of argument that the model is a true description of how the species evolved, what does it tell us about the appropriate definition of diversity?

Each of the species 1 through 6 consists of a long string of beads of identical length. It seems natural to define the difference or distance between any two species as the number of beads that are different between them. With this definition, the distance between any two species can be read from the corresponding genealogical tree of Figure 1.1 as the time back to their nearest common ancestor. The number of beads by which two species differ is equal to the time elapsed from their most recent common ancestor because that is exactly the time period over which the different beads have been independently accumulated by the two species.

One possible definition of diversity in the present context is the total number of *different* beads contained in the collection. Some reflection should reveal that, in the present model, diversity under this definition equals the length of the associated taxonomic tree. By the length of an evolutionary tree, I mean the total lengths of all its vertical branches, including the branch of the common ancestor of the entire family back to some unspecified outgroup. The reader should confirm that the number of different beads represented by the six existing species of Figure 1.1 is indeed equal to the total length of the associated tree under the bead model of an evolutionary branching process being assumed. Thus, under extreme simplifying assumptions about the nature of the underlying evolutionary process, if the diversity of a collection of species is defined to be the total number of different subunit-beads, then diversity equals the total branch length of the associated taxonomic tree.

There is an equivalent way of describing diversity in the above structure that is useful because part of it generalises. Without trying to be overly formal here, think heuristically of the operation of the bead process as a kind of "creation machine." Then the diversity of a collection is the number of "operations" required of the "creation machine" to make the collection. Later it will be shown how a more rigorous definition of the technology of a creation machine and its operations can be used to define diversity in the general case as the work required by the creation machine to make the diverse collection.

The preceding definition of diversity can be rephrased in terms of a "hierarchical search" procedure. In this context, think of a bead as a book. Then each species is like a library of M books. Any two libraries may contain certain books in common, and some that are different. Suppose every book embodied in the collection of species must periodically be searched to find, e.g., the appearance of a certain phrase. What is the best catalogue hierarchy for organising such periodic searches through all the books in the collection?

Some reflection reveals that the optimal hierarchical search structure is exactly the genealogical structure arising from the bead model that generates the evolutionary branching process. The highest level catalogue contains all books commonly held by every library $\{1, 2, 3, 4, 5, 6\}$. The next level of catalogue contains all books held in common by libraries $\{1, 2, 3, 4\}$, but not in $\{5\}$ and $\{6\}$. The next catalogue after that contains books in $\{1, 2, 3\}$ but not $\{4\}$. Then $\{1, 2\}$ but not $\{3\}$, then $\{1\}$ or $\{2\}$ but not both, then $\{5\}$ or $\{6\}$ but not both.

The optimal way to search every book is to first search the books in the highest catalogue, then the next highest, then the next highest after that, and so forth, until every book is searched. This hierarchical search procedure takes minimal time among all the alternatives because it completely avoids the redundancy of searching the same book twice in two different libraries or catalogues. The total time required to examine all the books in an optimal search hierarchy is here exactly the total branch length of the associated genealogical tree (equal to the total number of different books). Thus, an alternative definition of diversity is the minimal time for a complete hierarchical search. This definition is useful because it can form the basis for generalising the concept of diversity to a situation where the species, or libraries, do not conform to the bead model of an evolutionary branching process.

There is also a probabilistic interpretation, that can be motivated by a somewhat more operational criterion, but which yields the same essential identification of diversity with the total branch length of the associated taxonomic tree.

Suppose that we are looking for some desirable property, like a new source of food or medicine. Each collection of species can be viewed as a kind of natural portfolio of future options for the desirable property. Suppose that if the desirable property exists in a species, then it will be found in one or more of the beads out of which that species is composed. (Or, if the desirable property exists in a library, it will be found in at least one of the library's books.) Suppose that the probability of any particular bead having the desirable property is independent of any other bead having the desirable property and is equal to some small positive number ε. The probability that any bead does *not* contain the desirable property is then $\lambda = 1 - \varepsilon$.

Now we can calculate the probability that the entire collection of species does *not* contain the desirable property. Let L be the number of different beads in the collection of species, equal, as we have seen, to the total branch length of the corresponding taxonomic tree. Then the probability that the entire collection does not contain the desirable property is $P = \lambda^L$. A not unreasonable definition of the diversity of a collection of species might be the negative logarithm of the probability that the collection does not contain the desirable property. By this definition, diversity is kL, where $k \equiv -\log\lambda$ is a positive constant. Thus, either concept of diversity yields essentially the same construct – namely diversity equals the total branch length of the associated taxonomic tree.

The above reasoning gives a powerful way of thinking about the loss of diversity that accompanies extinction events, at least for the bead model of an evolutionary branching process that has been presented.

When any species becomes extinct, the loss of diversity equals the species' distance from its closest relative, and this myopic formula can be repeated indefinitely over any extinction pattern, because any subevolutionary tree of an evolutionary tree is also an evolutionary tree. When a species becomes extinct, the loss of diversity is calculated as if its evolutionary branch were snapped off the rest of the tree and discarded. This sharp mental image, properly used, permits a quick, exact visualisation of the effects of various combinations of species losses on diversity in the special case of perfect taxonomy based on the bead model of an evolutionary branching process.

A simple example may help to illustrate the basic issues. In Figure 1.1 is depicted a family tree representing the evolutionary history of six existing species. The two most closely related species are 5 and 6, so that the smallest loss from extinction of a single species occurs if one of these two vanishes. However, an analytical preservationist must be careful here. If, after species 5 goes extinct, species 6 also goes extinct, then the overall loss could be catastrophic since a whole evolutionary line will have been wiped out. While the diversity loss of 5 or of 6 is lower than that of any other single species in the collection, the diversity loss of the pair (5, 6) is greater than the diversity loss of any other pair in the set. Hence, an optimal conservation strategy might be to concentrate relatively few resources on saving species 5, if species 6 is reasonably safe, or it might involve concentrating relatively large resources on saving species 5, if species 6 has a high danger of extinction. I hope this kind of example, which could be repeated over a wide variety of different situations, illustrates the power of using the simple geometric interpretation of diversity as a conceptual aid for analysing policy options concerning preservation of diversity.

The previous reasoning can be pursued further to yield some not so obvious insights about conservation policy. Just to emphasise the abstract nature of the problem, suppose here we are talking about libraries. We have already mentioned how an expected diversity function can be defined when there is uncertainty. An expected diversity function is basically the sum of the deterministic diversity function of various collections of libraries, weighted by the existence probabilities of the various collections. Assume, as a simplification, that all survival probabilities are independent. Suppose the aim is to maximise expected diversity.

Consider the following numerical example. The numbers have been chosen to make the point sharply, but the point itself is quite general.

Referring again to Figure 1.1, let the survival probability of library 5 be $P_5 = .98$. Suppose the survival probability of library 6 is $P_6 = .02$. Library 5 might be called a relatively "safe" library, while library 6 is relatively "endangered."

Suppose now we consider the possibility of changing underlying resources to "shift" .01 of survival probability between libraries 5 and 6. Which of the following three alternatives yields the highest expected diversity?

Alternative	P_5	P_6
1. Status quo	.98	.02
2. Endangered library *more* endangered	.99	.01
3. Endangered library *less* endangered	.97	.03

I think it is fair to guess that most conservation-minded people would favour 3, the option that increases the survival probability of the endangered library at the expense of the safe library. Actually, expected diversity is *minimised* by alternative 3, while it is *maximised* by alternative 2.

The reason for this counterintuitive result can be explained as follows. For simplicity I just compare situations 2 and 3. The probability that *both* libraries *survive* is $(.99)(.01) = .0099$ in situation 2, while it is $(.97)(.03) = .0291$ in situation 3. This would seem to turn the calculation in favour of 3 and is probably what accounts for the intuition that 3 yields higher expected diversity than 2.

However, the probability that *both* libraries become *extinct* is $(.01)(.99) = .0099$ in situation 2, while it is $(.03)(.97) = .0291$ in situation 3, which is an exact reversal of the previous calculation.

Now it would, of course, be good for diversity to have both libraries 5 and 6 survive. But it would be a significant disaster for diversity if both libraries 5 and 6 went extinct, because a whole lineage of unique books would then have been extinguished. Therefore, other things being equal, the analytical preservationist favours making the safe library safer at the expense of making the endangered library more endangered, because a whole line may therefore be made safer – if a one to one tradeoff of survival probabilities is possible. Although this example rests upon specific assumptions, I believe it offers some relevant insights into conservation policy that could not easily be made outside the diversity function framework.

The perfect taxonomy structure induced by the bead model of an evolutionary branching process allows yet other powerful insights into the form of an optimal conservation policy. Consider, for example, the following idealised situation involving sharply posed preservation issues in such a context. This might be called the "Noah's ark problem."

Let the set S consist of n species denoted by $i = 1, 2, \ldots, n$. Let the (independent) probability that species i survives be denoted x_i. Each column n-vector $X \equiv (x_i)$ of survival probabilities defines an expected diversity function

$$U(X) \equiv E_x(V)$$

Suppose the objective function is of the form

$$\phi(X) = BX + U(X)$$

where b_i is the direct net benefit of species i and $B \equiv (b_i)$ is the row n-vector of direct net benefit coefficients.

Suppose the cost of preserving species i with probability x_i is equal to $c_i x_i$. (In the Noah's ark interpretation, c_i is the room in the ark taken up by the pair of species i.) Let the row n-vector of cost coefficients be $C \equiv (c_i)$. Let the total preservation budget be A. (A is the size of the ark.)

The simplest form of a constrained expected diversity maximising problem might be formulated as:

$$\text{maximise } \phi\,(X)$$

subject to:

$$CX \le A$$
$$0 \le x_i \le 1 \quad \text{for } i = 1, 2, \ldots, n$$

The previous constrained optimisation problem is well-defined, but it looks like a combinatoric nightmare. Actually, in the case of the bead model a simple myopic algorithm is available for solving the problem. It is here stated without proof.

The algorithm proceeds by eliminating the least valuable species, one species at a time, until the budget constraint is just met.

Suppose at some iteration the subset $Q \subseteq S$ of species exists with probability one, while the subset $S\backslash Q$ of species is extinct or exists with probability zero. ($S\backslash Q$ stands for the set S minus the set Q.) Suppose that the budget constraint is not being met:

$$\sum_{i \in Q} c_i x_i > A$$

The next step is to find the relatively least desirable species of Q. This is the species $j(Q) \in Q$ that satisfies the condition

$$\frac{b_j + d(j, Q\backslash j)}{c_j} = \min_{i \in Q}\left(\frac{b_i + d(i, Q\backslash i)}{c_i}\right)$$

Distance $d(j, Q)$ from point j to set Q is understood in the usual sense to be the distance from j to the element of Q closest to j.

The probability x_j is then brought down continuously from one towards zero until either the budget constraint is met or species j is eliminated, whichever occurs first. In the latter case, a new species set Q is defined which is equal to the previous species set Q minus the species j.[1] The procedure is repeated until the budget constraint is just met, at which point the algorithm has converged. The relevant theorem (not proved here) is that for the bead model of evolutionary branching such a myopic algorithm yields an optimal policy in the sense of satisfying the original optimisation problem. The theorem justifies using at each iteration a myopic benefit-cost ratio consisting of the traditional ratio of direct benefits to costs plus the diversity loss per preservation dollar. ,

The import of this approach consists in giving a rigorous global sig-

---

[1] Note that this will change some of the remaining $\{d(i, Q\backslash i)\}$ coefficients.

nificance to the strictly local decision-making index of species diversity loss per unit of conservation resources. Comparing "expected diversity loss per preservation dollar" among species thus turns out to be a legitimate extension of cost-benefit analysis.

I hope I have been able to present a fairly convincing argument that for the special bead model of an evolutionary branching process there is a moderately compelling case for identifying the diversity of a collection of species with the length of the associated genealogical tree. Unfortunately, the bead model is an extreme abstraction of an idealised evolutionary process. It provides a useful construct within which it really is fairly clear what we should mean by a diversity function. The difficult question, to which I next turn, is what to call a diversity function for a situation where the bead model is not strictly applicable.

Remember that the "distance" $d(i,j)$ between libraries i and j is the number of books different between i and j. In the bead model, all distances are "ultrametric," meaning that for any three libraries $i, j, k,$

$$\max \{d(i,j), d(i,k), d(j,k)\} = \text{mid} \{d(i,j), d(i,k), d(j,k)\}$$

Ultrametric distances have the enormously attractive property that they can be completely represented by a tree structure. Conversely, any (rooted, directed) tree defines a set of distances that are ultrametric.

This can be seen readily from Figure 1.1. The "distance" between any two species is represented in Figure 1.1 as the time back to the most recent common ancestor. Equivalently, this distance represents the collection of beads or books that are different between the two species or libraries.

I have tried to argue in this section that when distances are ultrametric, the leading candidate for a diversity function really should be fairly clear. With ultrametric distances, diversity is the total branch length of the associated tree. A variety of approaches or views support this interpretation. Furthermore, when distances are ultrametric a rather powerful theory can be developed to give insight into the nature of strategies that would maximise expected diversity. At this point we must address the issue of defining a diversity function in the more general case when distances are not ultrametric. We will proceed by attempting to generalise from the ultrametric case.

1.4   Diversity in the general case

Suppose we continue to think of a species as a collection of M beads on a string or M books in a library. Only now, the collection is not necessarily derived from the bead model of an evolutionary branching process, or distances are not necessarily ultrametric. How are we then to define the diversity of a group of species?

As before, the distance between species i and j is the number of beads or books that are different between them. Here the distances $\{d(i,j)\}$ are taken as given data. As is traditional, the distance $d(j,Q)$ from the point j to the set Q is equal to the distance from j to the element of Q closest to j.

It is only rarely that distances are ultrametric. Far more frequently, distances are not exactly as if they are derived from the bead model. We are typically confronted initially with a situation where the given pairwise symmetric dissimilarity-distance measures are not ultrametric and are therefore not consistent with the bead model of an evolutionary branching process.

In the general case of arbitrary distances, *the diversity function $V(S)$ is inductively defined to be the solution of the recursion*

$$V(S) = \max_{i \in S} \{V(S\backslash i) + d(i, S\backslash i)\} \qquad (1\text{-}1)$$

The dynamic programming Equation (1-1) is the centrepiece of the present approach to diversity. The solution of Equation (1-1) is unique once the initial conditions

$$V(i) \equiv d_0 \qquad \forall i \qquad (1\text{-}2)$$

are specified for any d_0. Depending upon the particular application, it is typically most convenient to normalise d_0 by setting it equal either to zero or to some large constant.

There are several possible axiomatic approaches that can be used to justify the diversity function of Equation (1-1). These axiomatic treatments are suggestively motivating, as I hope to indicate. However, the real argument for the diversity measure being proposed here is that it "works" fairly well – in the sense of creating a useful and consistent conceptual framework, while other measures "do not work" – in the sense that they violate one or more essential properties that a plausible diversity function should possess. The following condition seems like a basic axiom that is reasonable to impose on any diversity function.

Monotonicity in Species. If species j is added to collection Q, then

$$V(Q \cup j) \geq V(Q) + d(j, Q) \qquad \forall Q \qquad \forall j \notin Q \qquad (1\text{-}3)$$

where $d(j, Q)$ is the familiar (minimal) distance from point j to set Q.

The monotonicity in species Condition (1-3) expresses the intuitively desirable idea that the addition of any species to a group of species should increase diversity by at least the dissimilarity of that species from its closest relative among the already existing group of species. Or, conversely, monotonicity in species means that the extinction of any species of an ensemble causes a decline in diversity by no less than the distance of the extinguished species from its nearest neighbour in the ensemble.

Monotonicity in species is a "loose" property in the sense that it does not at all define a unique function because many diversity measures can be made to satisfy the Inequality (1-3). There are at least two ways to add a supplementary condition that would make the Inequality (1-3) hold so "tight" that it yields, in effect, the dynamic programming Equation (1-1).

The first approach is the most direct. View Condition (1-3) as a potentially very large set of constraints that must hold for *all* Q and for *all j*. Impose the uniform initialising Condition (1-2). Then simply define the diversity of S to be the minimum possible $V(S)$ that satisfies Equations (1-3), (1-2).

The reason this direct approach yields the dynamic programming recursion (1-1) is as follows. Suppose, by induction, the diversity functions $\{V(S\backslash i)\}$ have been defined for all i belonging to S. Then the smallest possible value for the diversity of S that would be consistent with Condition (1-3) must satisfy the condition:

$$V(S) \equiv \text{minimum } V \tag{1-4}$$

subject to:

$$V \geq V(S\backslash i) + d(i, S\backslash i) \qquad \forall \, i \in S \tag{1-5}$$

It is straightforward to confirm that the solution of Equations (1-4), (1-5) is Equation (1-1), which both proves the assertion and continues the induction argument to the next stage.

The problem of finding the smallest possible diversity function consistent with Equation (1-3) can be recast as an insightful evolutionary metaphor.

In this interpretation, the distance $d(i,j)$ stands for the number of (possibly weighted) character-state differences between i and j. For any set Q of existing species, $V(Q)$ here stands for the evolutionary length of Q, meaning the total number of character-state changes required to explain the evolution of Q under some rooted directed branching representation of the evolutionary process. For each species, the length from root to twig tip in this branching process is the same number M. Suppose that species $j \notin Q$ is added to Q to form the new set $Q \cup j$ of existing species. The number of *extra* character-state changes required to explain the evolution of j is at least the difference in character-state changes between j and its closest relative in Q, which is $d(j, Q)$. If j is added to Q, then at least $d(j, Q)$ additional character-state changes need to be explained. Therefore, any properly scaled feasible measure of evolutionary distance should simultaneously satisfy, for *all* Q and for *all j*, the basic consistency conditions of Equation (1-3).

It seems natural to define the diversity of S, denoted $V(S)$, to be the length of the tightest or most parsimonious feasible reconstruction of S, in the sense of being the minimal number of character-state changes required to account for the evolution of S. By the same argument as before, $V(S)$ so defined must satisfy Equations (1-4), (1-5), and, by extension, Equation (1-1). Thus, Equation (1-1) has the interpretation of describing the number of steps required to generate the most parsimonious "minimal evolution" branching structure that gives rise to the species of S.

This evolutionary metaphor can be recast as an insightful story about the cost of "making" diversity. Suppose the n objects each consist of M spaces or positions. Every position is filled with a particular colour, letter, flavour, codon, symbol, or whatever, depending on the context. Each object is in effect a mosaic of symbols. The "distance" $d(i,j)$ between objects i and j is the number of positions of i and j which have different symbols in them.

Think of a symbol in a particular position as being produced by a "stamping" or "punching" operation like a train conductor's hand puncher. When the tickets are lined up properly, it is just as easy to punch two or more tickets with the same symbol in the same position as it is to punch one ticket with that symbol.

More formally, symbols are stamped in place by a "creation machine" – some generalisation of the conductor's hand punching machine. At any intermediate stage of its manufacture, an object consists of completed stamped and uncompleted unstamped positions. The creation machine exhibits perfect economies of scale when identical stamping operations are performed on the same unstamped position of identical objects. Each identical operation on identical objects counts as only *one* operation. The appropriate image is that identical objects can be costlessly aligned so that the identical symbol may be punched in the same uncompleted position in one operation.

However, if the creation machine performs different operations on the same object or the same operation on different objects, no economies of scale are allowed and production is linear. In such cases the objects cannot be properly stacked up and more than one punching operation is required. The same operation performed on two different objects counts as two operations, just as do two different operations performed on two identical objects.

Now it seems natural to define diversity as the minimum cost of making the n objects different, as measured by the minimum total number of stamping or punching operations required by the creation machine. Essentially, diversity is the amount of work that the creation machine must do to create the different objects.

In the case of ultrametric distances, it is easy to see that diversity equals the total branch length of the corresponding tree. In the more general case, it is possible to derive a lower bound.

The bound is derived as follows. Suppose that Q ($\subset S$) is any set of species. Let the minimum number of operations required by the creation machine to make the collection Q be denoted $V(Q)$. Let j be a species in S but not in Q. Some reflection will reveal that the Inequality (1-3) must hold. Suppose that i is the species of Q that is closest to j. If species j is added to the collection Q, the very luckiest we might be in terms of minimising the number of operations on the creation machine is if all the positions of i that have different symbols from j happen to occur at the very end of the manufacturing sequence on the creation machine that made the collection Q, which includes i. In this fortuitous case, Equation (1-3) would hold with full equality. In the more general case, Equation (1-3) would hold as the stated inequality condition.

Now, the most optimistic number of operations of the creation machine required to make the collection S, consistent with the given distance data, must satisfy Equation (1-1). Thus, the function defined by Equation (1-1) has the interpretation of representing the minimal number of operations needed to create the diversity of the collection S.

There is yet another way of restating the evolutionary metaphor in terms of a bound on the search time for an optimal hierarchy. Here, think of each species as a library containing M books. The distance $d(i,j)$ is the number of books in library i but not in library j, or vice versa. Regularly, say once a week, every book must be searched to see if it contains some particular message, phrase, reference, formula, or whatever. It takes the same amount of time to search each book. Some books are common to two or more libraries, and the director wants to minimise the redundancy involved in searching the same book more than once. Suppose that search must be hierarchical, meaning that the catalogue must have a tree structure analogous to what is depicted in Figure 1.1. The highest level catalogue contains books held by every library. Then a bifurcation occurs which partitions the set of all libraries into two mutually exclusive subsets. The next level of (two) catalogues contains books held by all the libraries of one subset, but not by all the libraries of the other subset. Then further bifurcations occur which divide a subset of libraries into two mutually exclusive sub-subsets. Each sub-subset contains books held by all the libraries of one of the sub-subsets, but not by all the libraries of the other bifurcated sub-subset. This hierarchical catalogue process continues until all books in all libraries have been included.

In the bead model, the optimal hierarchical search procedure is identical to the genealogical tree, and it is perfectly effective in the sense that it completely avoids any redundancy of searching the same book twice in two different libraries or catalogues. When the bead model does not hold, some redundancy in hierarchical search is unavoidable. The question then becomes: what is the most economical search hierarchy in the sense of minimising total search time. Intuitively, one wants to group together libraries having a relatively large number of books in common to avoid redundant search.

When posed in full generality, it is impossible to find an optimal hierarchy without detailed information about which books are contained in every library. However, relying only on distance information does allow a theoretical lower bound on search time. This theoretical lower bound on total search time in an optimal hierarchy is what we will call the diversity of the collection of libraries.

The bound is derived as follows. Suppose that Q ($\subset S$) is any set of libraries. Let the optimal hierarchical search time for the collection Q be denoted $V(Q)$ and the optimal search tree for Q be denoted $T(Q)$. Let j be a library in S but not in Q. Some reflection will reveal that the Inequality (1-3) must hold. Suppose that i is the library of Q that is closest to j. If library j is added to the collection Q, the very luckiest we might be in terms of minimising hierarchical search time is if all the books of i that are different from j happen to occur at the very end of the part of the search tree $T(Q)$ that involves library i. Then the optimal search tree $T(Q \cup j)$ would look just like the optimal search tree $T(Q)$ except that an extra branch of length $d(i,j)$, which contains library j as an end twig, has been appended to the branch containing library i as an end twig. In this fortuitous case, Equation (1-3) would hold with full equality. In the more general case, Equation (1-3) would hold as the stated inequality condition.

Now, following the previous logic, the most optimistic hierarchical search time consistent with the given distance data must satisfy Equation (1-1). Thus, the diversity function of this paper has the interpretation of representing the minimal amount of time needed to perform an optimal hierarchical search of all the volumes contained in a given collection of species-libraries.

Another route to forcing the Inequality (1-3) to hold so "tight" that it yields the dynamic programming Equation (1-1) is to add an extra axiom to Equation (1-3) called the "link property." This new condition can be stated as follows.

Link Property: For all S, there exists at least one species $j(S) \in S$, called the "link" species, that satisfies

$$V(S) = d(j, S\backslash j) + V(S\backslash j) \tag{1-6}$$

As was shown in the last section, an especially appealing theoretical structure emerges in the case of bead model distances, where, in effect, Equation (1-6) holds for *all* $j \in S$. Unfortunately, it is mathematically impossible that Equation (1-6) can be true for all $j \in S$ in the general case of non-ultrametric distances. But from the link property it will at least be true always that the elimination of *some* species $j(S)$ will reduce diversity by exactly the distance of that species from its closest relative. The link property provides at least one tight natural connection between the derived value of diversity measure for any set and the primary distance data on which it is based.

That Conditions (1-3) and (1-6) imply Condition (1-1) is a fairly straightforward argument. There is also a probabilistic way of motivating the basic dynamic programming recursion of Equation (1-1) that deserves to be treated here. One of the most commonly cited reasons for maximising expected biodiversity is to maintain a kind of natural "portfolio diversification" of future options for finding new sources of food, medicine, and so forth. We will show that, under a not too bizarre model, the concept of diversity as "portfolio diversification" is really the same as the concept of diversity embodied by the diversity function previously defined.

Suppose, for concreteness, we are speaking of finding a pharmacological cure for some disease. If a species contains a cure, it will only become revealed over time, in the future. Thus, when a species becomes extinct the chance is lost forever that the species may be of later help in providing a medicine for treating the disease. What should we be preserving in such a context?

In this interpretation, let

$$P(i, j) \tag{1-7}$$

stand for an upper bound on the probability that species i does *not* contain a cure for the disease given that species j does *not* contain a cure. The data represented by Equation (1-7) are the basic, given, reduced-form primitives of the model.

It is assumed that the given conditional probability coefficients of Equation (1-7) are symmetric; for all i and j belonging to S,

$$P(i, j) = P(j, i)$$

Suppose the n species of S are produced by a process of "descent with modification" down an evolutionary tree, only we do not necessarily know the structure of the evolutionary tree.

In what follows, assume any particular evolutionary branching tree structure T out of all possible rooted directed trees that yield the species of S as labelled twig tip end-nodes of the evolutionary process. Each possible tree T defines a set of ancestor interior-nodes $A_T(S)$, ($n - 1$ of them in the bifurcating case) located within the branching structure.

Think of evolution as a branching process that results in the accretion of many tiny boxes. When two new species bifurcate from an ancestor node, they keep all the same tiny boxes they shared in common to that point, but henceforth they begin independently accreting different tiny boxes. If there is a pharmacological treatment for the disease, it will be found in one of the tiny boxes that has been accumulated along the evolutionary tree. In this model, the key structural assumption is that once a cure is contained in a parent node, then it is fixed or locked into all of the subsequent offspring nodes.

All statements that follow are with respect to the particular branching structure T being assumed. In other words, for convenience we are dropping the subscript T from the notation that follows, understanding that it is implicitly there.

Let Q and j satisfy Equations (1-4), (1-5). Let $P(j|Q)$ be the conditional probability that species j does *not* contain a cure given that each species of the set Q does *not* contain a cure. Let $A(Q)$ represent the set of all ancestor nodes of the set of species Q. Then,

$$P(j|Q) = P[j|A(Q)] \qquad (1\text{-}8)$$

The only way that the fact that Q does not contain a cure transmits information relevant to whether or not j contains a cure is through the knowledge that the ancestor nodes $A(Q)$ could not have contained a cure.

Let $a(j, Q)$ stand for the most immediate ancestor of j in the set $A(Q)$. Then,

$$P(j|A(Q)) = P(j|a(j, Q)) \qquad (1\text{-}9)$$

The entire relevance for the probability that j does not contain a cure given that $A(Q)$ does not contain a cure is summarised by the information that $a(j, Q)$, the most recent ancestor of j in $A(Q)$, does not contain a cure.

Applying basic probability theory to this special structure,

$$P(j|i) = P(j|a(j, Q)) \cdot P(a(j, Q)|i)) \qquad \forall i \in Q \qquad (1\text{-}10)$$

Taking the maximum of both sides of Expression (1-10) over all $i \in Q$ yields

$$\max_{i \in Q} P(j|i) = P(j|a(j, Q)) \cdot \max_{i \in Q} P(a(j, Q)|i)) \qquad (1\text{-}11)$$

Now $a(j, Q)$ ($\in A(Q)$) must be an ancestor node for *some* (at least one) $k \in Q$, implying that

$$P(a(j,Q)|k) = 1 \quad \text{for some } k \in Q \tag{1-12}$$

But then Equation (1-12) implies

$$\max_{i \in Q} P(a(j,Q)|i)) = 1 \tag{1-13}$$

Combining Equation (1-13) with (1-11) with (1-9) with (1-8), and using Definition (1-7) yields

$$P(j|Q) \le P(j,Q) \tag{1-14}$$

where

$$P(j,Q) \equiv \max_{i \in Q} P(j,i) \tag{1-15}$$

By definition

$$P(j|Q) \equiv \frac{P(Q \cup j)}{P(Q)} \tag{1-16}$$

where $P(Q)$ stands for the probability that none of the species of Q contain a cure for the disease. Combining Equation (1-14) with (1-16) yields

$$P(Q \cup j) \le P(Q) \cdot P(j,Q) \quad \forall Q \subset S \quad \forall j \in S \backslash Q \tag{1-17}$$

Let $\Pi(S)$ stand for the *maximum* value of $P(S)$ under the set of constraints signified by Equation (1-17):

$$\Pi(S) \equiv \max P(S) \quad \text{subject to Equation (1-17)} \tag{1-18}$$

If $\Pi(S \backslash i)$ were known for all $i \in S$, then Equation (1-17) (with $j \equiv i$, $Q \equiv S \backslash i$) implies that $\Pi(S)$ defined by Equation (1-18) must satisfy the dynamic programming recursion

$$\Pi(S) = \min_{i \in S} \{\Pi(S \backslash i) \cdot P(i, SP(i, S \backslash i)\} \tag{1-19}$$

It is convenient to transform (1-19) into an equivalent dynamic programming equation that is additive in distances by taking negative logarithms of all probabilities. Let

$$V(S) \equiv -\log \Pi(S) \tag{1-20}$$
$$d(i,j) \equiv -\log P(i,j) \tag{1-21}$$

Combining Equation (1-15) with (1-6) with (1-21) yields

$$d(i, S \backslash i) = -\log P(i, S \backslash i) \tag{1-22}$$

Using Equations (1-20), (1-21), (1-22), Equation (1-19) becomes transformed into the equivalent dynamic programming Equation (1-1).

From the isomorphism of Equation (1-19) with (1-1), the following important conclusion emerges. When the diversity function defined by (1-1) is being maximised, there is a well defined sense in which the worst-case probability of not being able to find a cure for the disease is simultaneously being minimised. Thus, provided that the distances are appropriately defined, maximising diversity is equivalent to minimising the worst-case risk of not being able to avoid some bad outcome in a portfolio choice problem. In other words, there is a not extremely bizarre model in which the two concepts of diversity are really two different sides of the same coin.

1.5   A possible application to Nature Conservancy rankings

The Nature Conservancy is a private nonprofit U.S. organisation dedicated to preserving rare or endangered species or natural communities by land acquisition programmes. In carrying out such a programme, the organisation requires operational criteria for determining site priorities.

Without going into full details, the Nature Conservancy approach to ranking the biodiversity potential of sites is very roughly the following. In Nature Conservancy methodology, the underlying "mapping units" or "elements" (species or communities) are ranked by how rare they are as measured by numbers of site occurrences: from G1 = critically important (5 or fewer occurrences) to G5 = demonstrably secure (over 100 occurrences). Then individual sites containing an element are graded by the likelihood that the element would survive on that preserved site: from A = highly likely to D = very unlikely. These two factors are then combined by prescribed guidelines to yield an overall biodiversity ranking of sites – from B1 = outstanding significance (e.g., presence of an A-ranked G1 element) to B4 = moderate significance (e.g., presence of a C-ranked G3 element). This ranking system is then used by the Nature Conservancy to prioritise the desirability of acquiring various sites.

The following question arises naturally. What is the relationship, if any, between the Nature Conservancy methodology and the diversity function approach described in this paper?

Under a number of simplifying assumptions, the Nature Conservancy biodiversity rank of a site can be interpreted as a rough approximation of the expected increase in biodiversity if the site were preserved. This may be shown as follows.

Suppose, for the sake of argument, the "elements" are species. Assume further that the distance from any species to any other species is one. (The diversity loss of each species is the same.)

Next, suppose a species i occurs on n sites. If any of these n sites is not protected, the probability of species i becoming extinct on that one site is

q. Assuming independence, the probability of species i becoming extinct in general (with loss of diversity one) is q^n.

Now suppose that one of the sites can be protected. Suppose this protection lowers the probability of extinction of species i on that site from q to p ($< q$). Then the probability that species i goes extinct in general becomes

$$pq^{n-1}$$

Defining

$$\delta \equiv q - p$$

the increase in expected diversity from protecting site i is

$$\Delta D = \delta q^{n-1}$$

In this special case, then, the change in expected diversity can be written as some function

$$f(\delta, n)$$

where f is an increasing function of δ and a decreasing function of n. In the above formula, δ is a measure of the change in site-specific survival probability or the degree of additional protection that would be given to the survival of species i on the site under consideration if that site were to be preserved.

The formula above provides a very rough justification of the Nature Conservancy ranking system. A site obtains a higher biodiversity ranking if it contains a more endangered species (n is smaller) and/or if the survival probability of the endangered species on that site is more greatly improved when the site is preserved (δ is bigger). The Nature Conservancy biodiversity ranking system corresponds very roughly to identifying sites whose preservation would cause a relatively large change in expected diversity. The Nature Conservancy methodology and the theory of diversity functions described in this paper thus dovetail quite nicely – at least under the greatly simplifying "special case" assumptions described above.

1.6   Summary and conclusions

In this paper I have tried to argue that if we are to get a handle on the conceptually very difficult problem of maximising diversity, we must be prepared to define a diversity function. Attempts to define diversity functions are in a beginning stage. I have tried to indicate one philosophy and one approach to defining diversity functions. This is not the place to develop fully the mathematical properties or the possible applications of the diversity function being proposed.[2] Nor is there sufficient space to make a complete comparison between this approach and others that have been proposed in the literature.[3] However, the reader should come away with at least some appreciation of the basic issues involved in defining appropriately a diversity function, and some sense of how such a definition might be constructed.

[2] To some extent these tasks have been attempted in my two other papers on this subject, Weitzman (1992, 1993).

[3] For some examples of other approaches to diversity, see the appropriate papers cited as references.

CHAPTER 2

Biodiversity in the functioning of ecosystems: an ecological synthesis

C. S. Holling, D. W. Schindler, Brian W. Walker and Jonathan Roughgarden[1]

2.1   Introduction

Analyses of biodiversity typically revolve around four questions, two of which are generally characterised as questions in ecology, and two as questions in economics. They are:

- How do natural processes and human actions affect the number and persistence of species?
- What different roles does biodiversity serve in determining the structure and function of ecosystems?
- What are the social and economic driving forces that generate human impacts on biodiversity?
- What is the value of biodiversity for human agricultural, ethical, medical, renewable resource and social purposes?

This chapter is written by ecologists for nonecologists – and primarily for economists. In one sense it is a primer describing our present state of understanding of ecosystems, the way they are structured, the way they function and the way they respond to disturbance and human management. In doing so, we concentrate on the first two questions. We attempt

[1] This paper owed its origin to the first meeting of the Biodiversity Project of the Beijer International Institute of Ecological Economics. We are indebted to the project leader, Charles Perrings and the Director of the Beijer Institute, Karl-Göran Mäler for making this possible. It was a meeting that brought together ecologists and economists. Early on in the meeting, each group met separately, and the synthesis that emerged from the ecologists' session formed the foundations for this paper. In addition to the authors, this included Robert Costanza, Carl Folke, W. H. Kemp, Ariel Lugo and Jacqueline McGlade, all of whom provided examples and contributed to the formalising of the synthesis. Each was willing to provide written contributions to the paper, but there proved to be inadequate time to do so. Had there been time, the paper would have been greatly enriched by their contributions and their experience with a wider range of examples of terrestrial and marine systems.

to do so in such a way that nonecologists can recognise concepts, explanations and descriptions that, if included in linked economic/ecological models and analyses, would produce an understanding of the problem that would be more consonant with present knowledge of the way ecosystems are structured and function.

Because the remaining two topics are the consequence of social, institutional and economic systems, we will leave those to our economist colleagues in the hope that they, in turn, will educate us in understanding how human dynamics intersect with nature.

For the first topic, we review the state of knowledge of the biotic, abiotic and human processes that affect diversity at different scales. We use examples from a number of ecosystems – boreal forest, fresh water, savanna and marine. For the second topic, we use recent advances in ecology to propose that different kinds of biotic diversity perform three distinct roles in determining the structure and function of ecosystems.

Much of what we say is based on models and understanding of ecosystems that are different from those that traditionally have been used by resource economists. These models and the underlying understanding have emerged from two streams of ecology that for a long time had remained separate – community ecology and ecosystem ecology. Levin (1992) suggests that the gulf between the two was the consequence of the different historical traditions in each. The need to merge community and ecosystem ecology and how to do it has been explored by an increasing number of ecologists (e.g., Holling 1992a; Schindler 1988, 1990a) and the issue was the subject of the recent 1993 Cary conference (Schindler 1994).

Community ecology emerged from basic studies of the interrelations between species, where generalised patterns were sought in those interactions among organisms. The abiotic physical environment was placed very much in the background, and was implicitly assumed to provide a constant template affecting organisms but not being affected by them. Both David Lack and Robert MacArthur called attention to coexisting bird species that "partitioned resources," which means that the various coexisting bird species either consume different types of food from the same type of place, or consume similar types of food from different kinds of places. When this observation is combined with the earlier laboratory and mathematical models showing that coexistence of competing species requires that they occupy different "niches," a general pattern was conjectured. That is, perhaps many, even all, ecological communities possessed a "structure" whereby coexisting species were each sufficiently different from one another. That is, species differences in "niche space" might resemble the interatomic distances within a crystal in physical space.

Alas, this search for an elegant pattern inherent in all ecological communities was frustrated by further work showing that readily discernible niche differences that could be related to competition between species, though not rare, were not common either, and could not usefully serve as a basis for a general theory of how ecological communities are organised. A similar fate has befallen all other proposed universal ecological generalisations to date, even as particular ecosystems have become increasingly well understood. As we shall see later in the context of marine systems, a consensus has developed on how an increasing variety of ecosystems function, though this has not yielded universal generalisations about how all ecosystems function. This does not preclude identification of the existence of common features to ecosystems, provided it is appreciated that no theoretical necessity is associated with such empirical generalisations.

In contrast to community ecology, many of the early approaches to understanding ecosystem dynamics emerged from applied studies of specific managed or mismanaged ecosystems, from studies of energy flow and transformation and of nutrient cycling. The applied studies were often more complete in their representation not only of biotic processes, but also of abiotic processes and of human policies and management. However, they suffered by producing a disconnected set of independent studies from which generalisation was difficult. There are now a few efforts that do generalise and synthesise the results of such studies – some from the perspective of human mediated impacts on ecological and environmental processes (Minns et al. 1990; Odum 1985; Schindler 1988, 1990b), and others from the perspective of the interrelation between policy, management and ecosystem structure and function (Holling 1986; Walters 1986).

The ecosystem energy and nutrient studies sought for more generality because they were rooted in knowledge of basic biogeochemical and physical processes. But they in turn suffered because they were often done in isolation from knowledge of the species in an ecosystem or from any detailed understanding of the manner in which species interacted. While specific studies were often complete, the lack of detailed information on community composition and species interaction made it difficult to generalise and extrapolate to ecosystems of different types. In addition, important concepts such as long-term changes and steady-state conditions were not usually addressed.

As a consequence of those differences, and not unexpectedly, practitioners in the different fields and approaches have often criticised those in the others. There are, for example, insufficient studies of ecosystems that have a duration and area of sufficient extent and resolution to provide confident generalisations about cross-scale interactions – from small to big and fast to slow phenomena. The importance of those cross-scale interactions have been recently highlighted by Levin (1992), Holling

(1992a), Likens (1992) and Fee et al. (1994). Moreover, some models of these ecosystems can be legitimately criticised for presuming that ever increasing detail improves prediction, although such examples are not at the heart of advances in the field (Walters 1986).

Similarly, premature theory has often implied generality when later studies have shown that the theories had rather narrow applications. After several decades of study, nature is still sufficiently surprising that, in principle, a deductive approach to theory is only useful in providing simple metaphors of how nature does not operate. Or at least we are still not smart enough to deduce important ecological patterns and structures and how they change under perturbations of various sorts, without a framework of thought and some case studies, however incomplete, to point the directions. Unlike neo-classical economics, effective ecological theory has proved to be inductive in its initial emphasis, not deductive.

Both community and ecosystem ecology have now matured sufficiently, however, that the differences between them have begun to blur. Community and population biologists have cooperated to address applied problems from a more dynamic perspective – e.g., of multispecies fisheries (May et al. 1979) and conservation (Soulé and Wilcox 1980). Other studies involving multidisciplinary teams have simultaneously studied both communities and ecosystem processes, or have examined effects of perturbations such as acid precipitation, eutrophication and engineered constructions on both community structure and ecosystem function (Likens 1992; Schindler 1987, 1988, 1990b; Schindler et al. 1991). The sub-disciplines have recently joined forces in grappling with the challenge of continental scale shifts in climate and land use, reflected, for example, in Brown and Maurer's (1989) call for "macroecology." And ecosystem ecologists have become active in formulating theory by utilising applied studies to provide focus and relevance – e.g., deriving simplified theoretical abstractions from complex empirically based models (Ludwig et al. 1978), or developing and testing general theories for responses of disturbed and transformed ecosystems (Odum 1985; Schindler 1987, 1990b).

As a consequence of this accumulation of analyses of a number of specific natural, managed and disturbed ecosystems, syntheses have become possible that suggest that the diversity and complexity of ecological systems can be traced to a small set of biotic and abiotic, or physical processes, each operating over different scale ranges (Holling 1992a; Fee et al. 1993). It is now generally recognised that on land, at least, animals and plants can shape their ecosystems (McNaughton et al. 1988; Naiman 1988; Schindler 1977) and that species dynamics (and hence biotic diversity) can be more sensitive to ecosystem stress than are ecosystem processes (Schindler 1990b; Vitousek 1990). That is, an ecosystem under stress apparently keeps much of its functions even though species composition changes. One recent explanation for this resilience and robustness of function is based on evidence that a relatively few processes, having distinct frequencies in space and time, structure terrestrial ecosystems, entrain other variables and set the rhythm of ecosystem dynamics (Holling 1992a). That particular advance provides one testable framework to begin to identify the ways in which biodiversity is affected and the consequence for ecosystem function.

In addition, the vulnerability of particular key structuring processes is a function of the number of alternative species that can "take over" a particular function when an ecosystem is perturbed (Schindler 1988, 1990b) i.e., functional diversity determines resilience. The least resilient or the most sensitive components of food webs, energy flows and biogeochemical cycles appear to be those where the number of species carrying out that function is very small. So far, this theory of "functional redundancy" has only been tested experimentally in aquatic ecosystems (Schindler 1990a).

In the remaining sections of this paper, we will provide more detail describing this emerging synthesis and its consequence for defining priorities for conservation or restoration of biodiversity.

2.2   Features common to many ecosystems

The accumulated body of empirical evidence concerning natural, disturbed and managed ecosystems identifies key features of ecosystem structure and function that we suspect are not included in many economists' image of ecology, i.e.:

- Change is not continuous and gradual. Rather it is episodic, with slow accumulation of natural capital such as biomass or nutrients, punctuated by sudden releases and reorganisation of that capital as the result of internal or external natural processes or of man-imposed catastrophes. Rare events, such as hurricanes, or the arrival of invading species, can unpredictably shape structure at critical times or at locations of increased vulnerability. The results of these rare events can persist for very long periods. Irreversible or slowly reversible states exist – i.e., once the system flips into such a state, only explicit management intervention can return its previous self-sustaining state, and even then success is not assured (Walker 1981). *Critical processes function at radically different rates covering several orders of magnitude, and these rates cluster around a few dominant frequencies.*
- Spatial attributes are not uniform or scale invariant. Rather, pro-

ductivity and textures are patchy at all scales from the leaf, to the landscape, to the planet. There are several different ranges of scales each with different attributes of patchiness and texture (Holling 1992a). *Therefore scaling up from small to large cannot be a process of simple linear addition; nonlinear processes organise the shift from one range of scales to another.*

- Ecosystems do not have single equilibria with functions controlled to remain near it. Rather, destabilising forces far from equilibria, multiple equilibria and absence of equilibria define functionally different states, and movement between states maintains structure and diversity. *On the one hand, destabilising forces are important in maintaining diversity, resilience and opportunity. On the other hand, stabilising forces are important in maintaining productivity and biogeochemical cycles, and even when these features are perturbed, they recover rather rapidly* (e.g., recovery of lakes from eutrophication or acidification). (Schindler 1990a; Schindler et al. 1991).

- Policies and management that apply fixed rules for achieving constant yields (e.g., fixed carrying capacity of cattle or wildlife, or fixed sustainable yield of fish or wood), independent of scale, lead to systems that increasingly lack resilience i.e., to ones that suddenly break down in the face of disturbances that previously could be absorbed (Holling 1986). *Ecosystems are moving targets, with multiple futures that are uncertain and unpredictable. Therefore management has to be flexible, adaptive and experimental at scales compatible with the scales of critical ecosystem functions* (Walters 1986).

The features described above are the consequence of the stability properties of natural systems. In the ecological literature, these properties have been given focus through debates on the meaning and reality of the resilience of ecosystems. The resolution of these arguments about resilience underlies the more recent synthesis that has emerged concerning the way ecosystems are structured and function across scales in time and space. For that reason, and because the same debate seems to be emerging in economics, we will review the concepts in order to provide a foundation for the synthesis.

Resilience of a system has been defined in two very different ways in the ecological literature. These differences in definition reflect which of two different aspects of stability are emphasised. The consequences of those different aspects were first contrasted by Holling (1973) for ecological systems, in order to draw attention to the paradoxes between efficiency and persistence, or between constancy and change, or between order and disorder.

One definition, and the more traditional, concentrates on stability near an equilibrium steady-state, where resistance to disturbance and speed of return to the equilibrium is emphasised (Pimm 1984; O'Neill et al. 1986). That view provides one of the foundations for economic theory as well. We will term that resilience of the first kind.

The second definition emphasises conditions far from any equilibrium steady-state where instabilities can flip a system into another regime of behaviour – i.e., to another stability domain (Holling 1973). In this case the important measurement of resilience is the magnitude of disturbance that can be absorbed before the system changes its structure by changing the variables and processes that control behaviour. We will call that resilience of the second kind.

The same differences have also begun to emerge in economics with the identification of multistable states for competing technologies because of increasing returns to scale (Arthur 1990). Thus, increasingly it seems that effective and sustainable development of technology, environments and ecosystems requires ways to deal with not only near equilibrium efficiency but the reality of more than one equilibrium. If there is more than one equilibrium, in which direction should the finger on the invisible hand of Adam Smith point?

Each of these two aspects of a system's stability has very different consequences for evaluating, understanding and managing complexity and change. Here, and in other chapters of this book, the definition of resilience used is that of the second kind: the amount of disturbance that can be sustained before a change in system control or structure occurs. We do so because that interplay between stabilising and destabilising properties is at the heart of questions of persistence and hence of changes in biodiversity.

The two contrasting aspects of stability – essentially one that focuses on *efficiency* of function vs. one that focuses on *existence* of function – are so fundamental that they can become alternative paradigms whose devotees reflect traditions of a discipline or of an attitude more than of a reality of nature.

Those who emphasise the near equilibrium definition of ecological resilience (resilience of the first kind), for example, draw predominantly from traditions of deductive theory and modelling (Pimm 1984) where simplified, untouched ecological systems are imagined or from traditions of engineering (De Angelis 1980; O'Neill et al. 1986; Waide and Webster 1976), where the motive is to produce systems with a single operating objective. This both makes the mathematics more tractable and accommodates the engineer's goal to develop optimal designs. There is an implicit

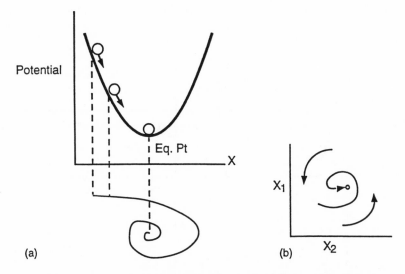

Figure 2.1. Two views of a single, globally stable equilibrium. (a) Provides a mechanical ball-and-topography analogy; (b) provides an abstract state space view of a point's movement towards the stable equilibrium, with x_1 and x_2 defining, for example, population densities of predator and prey, or of two competitors.

assumption that there is global stability – i.e., there is only one equilibrium steady-state – or at least that if other operating states exist, they are irrelevant (Figure 2.1).

In this example, resilience of the first kind is measured by the resistance of the ball to disturbances away from the equilibrium point and the speed of return to it.

Those who emphasise the stability domain definition of resilience (resilience of the second kind) come from traditions of applied mathematics and applied ecology at the scale of ecosystems – e.g., of the dynamics and management of fresh water systems (Fiering 1982), of forests (Holling et al. 1977), of fisheries (Walters 1986), of semi-arid grasslands (Walker et al. 1969) and of interacting populations in nature (Dublin et al. 1990; Sinclair et al. 1990). Because these studies are rooted in inductive theory formation and in experience with the impacts of large scale management disturbances, the reality of flips from one operating state to another cannot be avoided. Management and resource exploitation can overload waters with nutrients, turn forests into grasslands, trigger collapses in fisheries and transform savannas into shrub dominated semideserts (Figure 2.2).

One example, described elsewhere in this book (Perrings and Walker Chapter 6) concerns grazing of semiarid grasslands. Under natural condi-

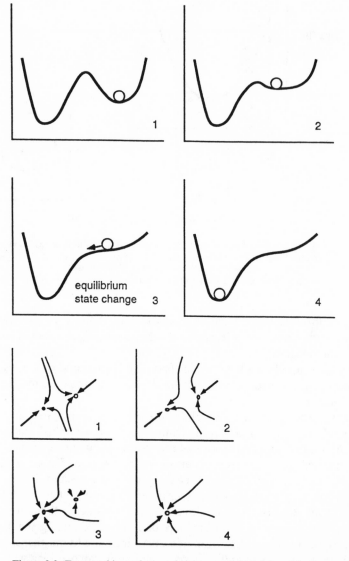

Figure 2.2. Topographic analogy and state space views of evolving nature. The system modifies its own possible states as it changes over time from 1 to 4. In this example, as time progresses, a progressively smaller perturbation is needed to change the equilibrium state of the system from one domain to the other, until the system spontaneously changes state. (a) Ball-and-topography analogy; (b) equivalent state space representation.

tions in east and south Africa, the grasslands were periodically pulsed by episodes of intense grazing by various species of large herbivores. Directly as a result, a dynamic balance was maintained between two groups of grasses. One group contains species able to withstand grazing pressure and drought because of deep roots. The other contains species that are more efficient in turning the sun's energy into plant material, are more attractive to grazers but are more susceptible to drought because of the concentration of biomass above ground in photosynthetically active foliage.

The latter, productive but drought sensitive grasses, have a competitive edge between bouts of grazing. But, because of pressure from pulses of intense grazing, that competitive edge for a time shifts to the drought resistant group of species. The result of these shifts in competitive advantage is that a diversity of grass species is maintained that serves a set of interrelated functions – productivity on the one hand and drought protection on the other.

When such grasslands are converted to cattle ranching, however, the cattle have been typically stocked at a sustained, moderate level, so that grazing shifts from the natural pattern of intense pulses separated by periods of recovery, to a more modest but persistent impact. The result is that the productive but drought sensitive grasses consistently have advantage over the drought resistant species and the soil and water holding capacity they protect. The species assemblage narrows to emphasise one functional type. Droughts that previously could be sustained then no longer can be and the system can suddenly flip to become dominated and controlled by woody shrubs. It is an example of what Schindler (1990a, 1994) has demonstrated in lakes as the effect of a reduction of functional diversity.

There are many examples of managed ecosystems that share this same feature of gradual loss of functional diversity with an attendant loss of resilience, followed by a shift into an irreversible state – e.g., in agriculture, forest, fish and grasslands management. In each case the cause is reduction of natural variability of critical structuring variables – e.g., plants, insect pests, forest fires, fish populations, or grazing pressure – with the result that the ecosystem evolves to become more spatially uniform, less functionally diverse and more sensitive to disturbances that otherwise could have been absorbed (Holling 1986). That is, resilience of the second kind decreases.

Moreover, such changes can be essentially irreversible because of accompanying changes in soils, hydrology, disturbance processes and keystone species complexes. Control of ecosystem function shifts from one set of interacting physical and biological processes to a different set.

In the face of such flips, near equilibrium behaviour and control seems irrelevant and the prescriptive goal shifts from questions of maximising constancy of yield to one of designing interrelations between people and resources that are sustainable in the face of surprises and the unexpected.

The heart of these two different views lies in assumptions of the existence of multistable states or not. If it is assumed that only one stable state exists or can be designed to so exist, then the only possible definition and measures for resilience are near equilibrium ones – such as characteristic return time. And that is certainly consistent with engineering's desires to make things work, not to intentionally make things that break down or suddenly shift their behaviour. But nature is different.

There are different stability domains in nature, so that a near equilibrium focus seems myopic and attention shifts to determining the constructive role of instability in maintaining diversity and persistence, and to designs of management that maintain ecosystem function in the face of unexpected disturbances – in short, that maintains or expands resilience of the second kind. It is those ecosystem functions and resilience of the second kind that provides the ecological "services" that invisibly provide the foundations for sustaining economic activity.

It is these views of stability and of functional diversity that provide a foundation for an emerging synthesis that has three elements, each of which we will deal with in turn in what follows. The first deals with ecosystem structure across scales from metres and days to thousands of kilometres and millennia. The second connects that structure with ecosystem functions and identifies different classes of biodiversity that serve different functions in ecosystems. The third explores the way human impacts change diversity.

2.3   Scales and structure: an example

An ecosystem consists of a biological community that interacts with the physical and chemical environment, with adjacent ecosystems and with the atmosphere. Functions include biogeochemical cycling, i.e., the conversion, recycling and transfer to other systems of critical chemicals that affect growth and production; and energy flow, the capture of solar energy by photosynthesis, the transfer of that energy between organisms, and its metabolism and storage. These two processes set an upper limit on the quantities and numbers of organisms and on the number of trophic or energy transforming and storing levels that can exist in an ecosystem, and hence indirectly on the biodiversity of the biological community.

Scales of observation and function determine both the perception of what diversity is and how it is generated and operates. A classic forest on the granitic shield country of eastern continental North America (often

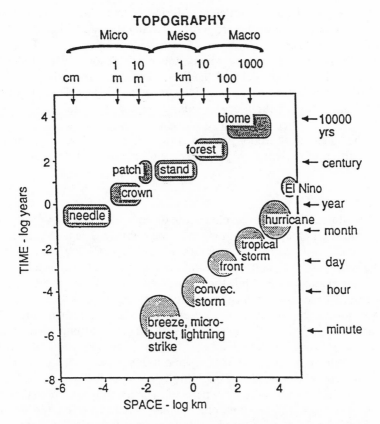

Figure 2.3. Space/time hierarchy of the boreal forest and of atmosphere (Holling, 1992a).

termed a boreal forest, because it is characteristic of the high latitude "boreal" region of the planet) is an excellent example. It is structured as a nested hierarchy, each level having its own scale ranges in both time and space. Figure 2.3 illustrates that, and adds the scales of atmospheric variation that couple with the ecosystem.

In the following section we will describe some of the key functions and properties that form each component of that boreal forest hierarchy, proceeding from the patch, to the stand, to the contiguous forest, to the continental boreal forest and finally to the circumpolar boreal biome. We could have equally chosen a savanna and grassland system like that in East Africa, or a wetlands system like that of the Everglades of Florida, or the temperate forests of central Europe. So long as these systems were terrestrial, the essential features and dynamics would be similar – only the specific species would be different. Marine systems, however, are different, as a consequence of the properties of water, but we will deal with those differences later.

To some readers there will be too much detail in what follows, but therein is where the devil of biodiversity lies. To others there will be too little detail, but in generality lies the understanding needed to form policy. We have tried, in our effort for balance between specificity and generality, to provide sufficient flavour of the specific reality of natural processes so that, in the next section we can turn to a more succinct summary and synthesis of present theory of ecosystem function and structure.

The patch

The dominant processes influencing diversity at the scale of the patch in a forest are competition among plant species for nutrients, light and water. As will be noted shortly, animal's use of plants for food and protection certainly affects plant composition and structure, but typically at scales larger than a patch. While the particular species of trees in a boreal forest vary between sites, typically only two or three tree species predominate on upland boreal sites. In the wetter eastern boreal for example, birch, spruce and balsam fir are typical, with birch flourishing early in succession, spruce in intermediate and balsam later. This sequence reflects a competitive ordering, where spruce out-competes birch and balsam out-competes both spruce and birch for light, in particular.

In the fire-influenced drier parts of the midcontinental boreal, however, poplar, birch and jackpine predominate. Typically, birch and poplar appear quickly after fire, because they resprout from the roots of fire-killed trees. Jackpine grow more slowly, and often do not predominate for several years after a fire. While balsam, fir, cedar and white spruce may appear on sites that are unburned for 150 years or more because jackpine typically die after 100–120 years, typical fire cycles of 100 years or less leave jackpine as a typical upland dominant.

The foundations for biochemical cycling and energy accumulation and storage are controlled at the patch scale by physiological processes of photosynthesis, respiration, decomposition and nutrient dynamics, which are reflections of the tree species, understory vegetation and microbial communities.

Patches can be a variety of sizes, reflecting combinations of fire size and intensity, soil characteristics, stand age, and meteorological conditions. They typically are about 30 metres across because that span covers the range of competitive influence of a single mature tree. The patch properties of texture and variety allow some species, like soil organisms, to live their whole lives within that patch, and others, like the hundreds of species of insects, to live all their immature stages at the patch scale. But many other animals, particularly the birds and mammals, operate on a larger scale, where stands of trees and the relation between stands becomes an important determinant of their existence and function.

The stand

Beyond the scale of a patch is a stand of trees that originated from a specific disturbance event. In the forest of the boreal region, the size of the stand can be up to 500 hectares or more. Its size and heterogeneity is determined by disturbance processes such as fire or insect outbreak. Such disturbances are contagious or spreading over space. Once they exceed a certain patch size, they become self-propagating and freed from control by the originating trigger – often weather. Such mesoscale processes transfer dynamics of a patch to the scale of stands.

The critical processes that control stand dynamics in the boreal forest are not biological processes of plant competition, as in the patch, but abiotic and zootic ones of disturbance – storm, fire, insect outbreak, herbivory by mammals and disease. Such processes can completely transform the outcome of plant competition and growth determined at the scale of the patch, and thereby maintain diversity that otherwise would be constrained. For example, although fir can out-compete spruce in many localities, it does not "take-over," because it is particularly vulnerable and sensitive to periodic attack by an insect defoliator called spruce budworm, which, long before people exploited the forest, regularly swept across the high latitude eastern half of the North American continent, killing up to 80% of the balsam in its wake. The outbreaks occurred in a quasicycle of periods ranging from 40 to 100 years and left behind areas of disturbance up to 500 hectares in size. Spruce is less vulnerable to budworm and birch not at all. Hence the insect disturbance pattern shifts the balance in competition from one coniferous species to another – much as the pulses of grazing in semiarid grasslands, described in the previous section, shift competition between functionally different groups of grasses.

Between outbreaks, the rapidly growing balsam could begin to dominate spruce. Outbreaks reversed that competitive edge, because balsam was so much more vulnerable to attack by budworm, allowing spruce to sustain itself until the surviving understory of balsam could again catch up. Hence the diversity of tree species is maintained by a combination of within-patch plant competition operating over tens of metres at yearly scales and budworm dynamics operating over hundreds of metres and decades.

The contiguous forest

At the next larger scale beyond a stand, is a contiguous forest. There the proportion of fir to spruce, for example, is correlated with the intensity of outbreaks and this in turn reflects differences in regional climate. The full geographical area of budworm-modified forest extends over a range of climatic regimes from continental in the west to maritime in the east. Spruce is more abundant in northwestern Ontario, both coniferous species are present in approximately equal numbers in New Brunswick and, further to the east, balsam fir dominates in Newfoundland. The climate in each of those regions differs – the warm dry springs that favour budworm survival are more common in the west, become less frequent in New Brunswick and still less so in the cool maritime climate of Newfoundland.

When these different climate conditions were introduced into a space/time simulation model of the forest and its dynamics (Clark et al. 1979), the simulated outbreaks followed the same progression, becoming progressively less severe from north Ontario, to New Brunswick, to Newfoundland. Thus the proportion of different tree species correlates well with intensity of outbreaks: the more intense the outbreak, the more spruce; the less intense, the more balsam fir. Differences in the frequency of warm dry springs thus affect tree species composition through the weather's influence on budworm as a disturbance process. This example is not at all atypical.

Another example is how the fire cycle appears to affect terrestrial-aquatic linkages in the boreal region. In old-growth areas, beaver are usually present but rather rare, unless heavily trapped, due to the scarcity of small deciduous trees which provide them food. For several years after a fire, however, small aspen and birch dominate the recovering forest. Beaver quickly multiply, removing almost all of these species in areas that are accessible from streams and lakes, feeding on the bark and building dams and lodges with the logs. As a result, coniferous species that grow much slower are eventually released, becoming dominant in the forests. Of course, beaver dams also dramatically alter stream flow patterns and lake levels, greatly changing the character of boreal landscapes and of biogeochemical processes (Naiman 1988). Fires also cause decreased losses of dissolved organic matter, which control the depth of the euphonic zone in boreal lakes, and cause increased mixing depths in small lakes due to greater wind exposure (Schindler et al. 1990; Schindler 1990b). Thus, the structure and function of terrestrial and aquatic boreal ecosystems are closely interlinked.

In summary, the structure, dynamics and diversity of terrestrial ecosystems is the consequence of an interaction between regional scale weather phenomena, stand level processes of zootic and abiotic disturbance and patch level processes of plant competition.

The continental boreal forest biome: spatial considerations

Finally, if we extend time still further to several lifetimes of trees and expand spatial perspective to a continental scale, it becomes clear the disturbance events are periodic. Such disturbances are not intrusions from outside but are an inherent part of ecosystem succession. In the case of a Douglas fir stand on the storm-swept west coast of Vancouver Island, or overmature jackpine forest in northwestern Ontario, the periodic disturbances could well be windstorms that can "clear-cut" many hundreds of hectares as a normal process when extremes of weather intersect with increasing vulnerability of aging trees. For different tree species, in different regions of the boreal forest, the natural disturbance might be an insect outbreak, windstorm or drought, which in turn are often preconditions for a forest fire (Heinselman 1973). Sustainability and diversity at that scale can be seen as the maintenance of successional cycles of stand level boom-and-bust to produce a perpetuating mosaic of stands of trees of different ages, each stand covering 100 to 1,000 hectares over the boreal forested region.

By this time and at this geographic extent, we are describing a good part of the present unlogged, unmanaged high latitude forest of North America and Eurasia – the boreal biome. At this scale, there are groups of ecosystems of coniferous and mixed forests and of bogs, wetlands, rivers and lakes, all interacting with each other. At this scale the stands can be of varied species, depending in large measure on soil and moisture conditions. Larch trees dominate lowland wet areas, the spruce/fir complex on moist uplands and jackpine on dry areas.

The interrelationships among these different boreal ecosystems types are easily recognised in the boreal zone. In particular, the long-term history of terrestrial, wetland and lacustrine ecosystems is often preserved in the laminated sediments of small lakes, or in peat bogs where sediment ages can be documented by radiometric dating. A number of studies have also documented the long-term interactions of geochemical cycles in uplands, wetlands, streams and lakes in the boreal (Bayley et al. 1993; Schindler et al. 1976, 1980, 1990).

The continental boreal forest biome: temporal considerations

These terrestrial and aquatic ecosystems aggregate to form the boreal forest biome whose existence is itself a passing and transient thing, something that emerged in its present form perhaps 8,000 years ago following the retreat of ice sheets. Pollen records demonstrate that the aggregations of tree species following the retreat of the ice sheets was a highly individualistic process depending on individual species' response to weather, their unique dispersal properties and the distance to sources of seeds. The processes defining the system at this scale now include geophysical cycles that are responsible for rhythms of glaciation, erosion and land movement. Now the duration of geological stages begin to determine diversity. For example, the ancient Lakes Baikal and Tanganyika have much greater biological diversity than lakes in North America at similar latitudes where productivity and nutrient chemistry became established only since the last glaciation 10,000 years ago. Spatial connectivity among a mosaic of ecosystems is equally important. The biota of islands and lakes far from similar formations are generally poorer than in continental systems or in regions with lakes close to each other (Schindler 1990a).

Let's pause, at this point, for a summary. At each of the hierarchical levels from patch, to stand, to forest, to biome, a different cluster of processes control ecosystem function. Plant and local biogeochemical processes dominate at small scales and climatic, geophysical and global biogeochemical processes are more important at large scales. At the intermediate scales, zootic and abiotic disturbance processes dominate. They include insects that defoliate and tunnel, ungulates that graze and browse, top carnivores that predate, fires, wind and disease. Until the development of satellite images and the software to analyse them, these processes operated on scales too large to observe directly. They sometimes seemed almost biblical in their function and could equally be called forces of plague, pestilence, fire and storm! Their interaction across hierarchical levels and temporal and spatial scales determines diversity.

How can we deal with such multiscale complexity when our best science and models are scale limited? One simplification is possible because there is a universal feature of the critical processes at each of the levels, over all the scales from needle to continent. Each functions as a cycle of birth, growth, death and renewal, but each with its own unique speed from months for needles to millennia for continental landscapes. What sustains such cycles? Oddly, it is the processes of death and renewal rather than those of birth and growth that lie at the heart of sustainability and diversity. That is where we need to search for measures of sustainability and to discover the functions of diversity – the capacity to renew after disturbance. Consider ecosystem succession.

2.4   Ecosystem functions and roles for diversity

Over the last decade, the literature on ecosystems has led to major revisions in the original Clementsian view of succession (Clements 1916). That initial view was one of a highly ordered sequence of species assemblages moving toward a sustained climax assemblage of species whose characteristics were determined by climate and soil conditions. That view or theory is not too different in its fundamentals from theories in neoclassical economics. Both assume single equilibrial states with regulatory (negative feedback forces) providing the "invisible hand" that guides the system along some optimal trajectory to an optimal state.

In ecology, the revision to this view comes from studies of nature itself:

- From extensive comparative field studies of a number of different ecosystems (Walker, 1988; e.g., West et al. 1981)
- From critical experimental manipulation of ecosystems that establish some of the alternative paths to succession (Bormann and Likens 1981; Schindler, 1988; Vitousek and Matson 1984) that establish some of the alternative paths to succession
- From paleoecological reconstructions that demonstrate multiple stable states and much more variable paths to those different states (Davis, 1986; Delcourt et al. 1983)
- From studies that link systems models and field research (Clark et al. 1979; West et al. 1981; Westoby et al. 1989)

The revisions include four principal points:

- Firstly, the species that invade after disturbance and during succession can be highly variable and are determined by the type, timing and intensity of chance events.
- Secondly, both early and late successional species can be present continuously and simultaneously.
- Thirdly, large and small disturbances triggered by events like fire, wind and herbivores are an inherent part of the internal dynamics and in many cases set the timing of successional cycles.
- Fourthly, some disturbances can carry the ecosystem into quite different stability domains – for example, fire can transform mixed grass and tree savannas into shrub dominated semideserts, for example (Walker 1981) – i.e., there is more than one possible "climax" state and each is controlled by a different ensemble of species under different physical conditions of soil.

In summary, therefore, the notion of a unique optimal path to a sustained optimal climax is a static and unrealistic view. The combination of these advances in ecosystem understanding with studies of population systems has led to one version of a synthesis that emphasises four primary stages in a terrestrial ecosystem cycle (Holling 1986).

The renewal cycle

The traditional view of ecosystem succession has been usefully seen as being controlled by two functions: *exploitation,* in which rapid colonisation of recently disturbed areas is emphasised and *conservation,* in which slow accumulation and storage of energy and material is emphasised. To an economist or organisation theorist those functions are roughly equal to the entrepreneurial market for the exploitation phase and the bureaucratic hierarchy for the conservation phase.

But the revisions in ecological understanding indicate that two additional functions are needed (Figure 2.4). One is that of *release,* or "creative destruction," a term borrowed from the economist Schumpeter (as reviewed in Elliott 1980), in which the tightly bound accumulation of biomass and nutrients (generally referred to as *aggradation*) becomes increasingly fragile (overconnected in systems terms) until it is suddenly released by agents such as forest fires, insect pests or intense pulses of grazing.

The second is one of *reorganisation,* in which soil processes minimise nutrient loss and reorganise nutrients so that they become available for the next phase of exploitation. This last phase is essentially equivalent to processes of innovation and restructuring in an industry or in a society – the kinds of economic processes and policies that come to practical attention at times of economic recession or social transformation.

During this cycle, biological time flows unevenly. The progression in the ecosystem cycle proceeds from the exploitation phase (Box 1, Figure 2.4) slowly to conservation (Box 2), very rapidly to release (Box 3), rapidly to reorganisation (Box 4) and rapidly back to exploitation. During the slow sequence from exploitation to conservation, connectedness and stability increase and a "capital" of nutrients and biomass is slowly accumulated. For an economic or social system the accumulating capital could as well be infrastructure capital, levels in the organisation or standard operating techniques that are incrementally refined and improved.

In forest ecosystems, as the progression to Box 2 proceeds, the nutrient capital becomes more and more tightly bound within existing vegetation, preventing other competitors from utilising the accumulated capital until the system eventually becomes so overconnected that rapid change is triggered. The agents of disturbance might be wind, fire, disease, insect outbreak or a combination of these. The stored capital is then suddenly released and the tight organisation is lost to allow the released capital to be reorganised to initiate the cycle again. Human enterprises can have similar behaviour, as, for example, when an industry like IBM or General Motors accumulates rigidities to the point of crisis, followed by efforts to restructure.

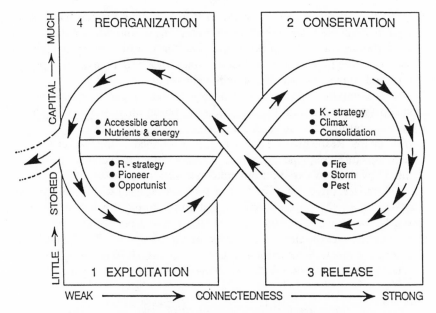

Figure 2.4. The four ecosystem functions and the flow of events between them. The arrows show the speed of that flow in the ecosystem cycle, where arrows close to each other indicate a rapidly changing situation and arrows far from each other indicate a slowly changing situation. The cycle reflects changes in two attributes: (1) the Y axis: the amount of accumulated capital (nutrients, carbon) stored in variables that are the dominant controlling variables at the moment; and (2) the X axis: degree of connectedness among variables. The exit from the cycle indicated at the left of the figure indicates the stage in which a flip is most likely into a less or more productive and organised system, i.e. devolution or evolution as revolution!

That pattern is discontinuous and is dependent on the existence of changing multistable states that trigger and organise the release and re-organisation functions. For example, the release role described in an earlier section for spruce budworm in the eastern balsam fir forest occurs because the maturing forest accumulates a volume of foliage that eventually dilutes the effectiveness of the searching capacity of insectivorous birds whose populations, in younger stands, control budworm populations. Essentially a lower equilibrium density for budworm is set by a "predator pit" in a stability landscape during the phase of slow regrowth of the forest (Clark et al. 1979; Holling 1988). This "pit" eventually collapses as the trees mature to release an insect outbreak and reveal the existence of a higher equilibrium. A similar argument can be described for release by fire, as a consequence of the slow accumulation of fuel as the forest ages.

To summarise and generalise this example, for long periods in a regrowing forest, the slow variable (trees) controls the fast and intermediate speed ones (budworm and foliage or fire and fuel) until a stability domain shrinks to the point where the fast variables for a brief time control behaviour and triggers release of accumulated capital.

It is instabilities and chaotic behaviour which trigger the release phase, which then proceeds in the reorganisation phase where stability begins to be reestablished. In short, chaos emerges from order, and order emerges from chaos! Resilience and recovery are determined by the fast release and reorganisation sequence, whereas stability and productivity are determined by the slow exploitation and conservation sequence.

Moreover, there is a nested set of such cycles, each with its own range of scales. In the typical boreal forest for example, fresh needles cycle yearly, the crown of foliage cycles with a decadal period and trees, gaps and stands cycle at close to a century or longer periods. The result is the ecosystem hierarchy described earlier, in which each level has its own distinct spatial and temporal attributes (as was suggested in Figure 2.1).

Evolutionary change and the adaptive cycle

A critical feature of such hierarchies is the asymmetric interactions between levels (Allen and Starr 1982; O'Neill et al. 1986). In particular, the larger, slower levels maintain constraints within which faster levels operate. In that sense, therefore, slower levels control faster ones. If that was the only asymmetry, however, then hierarchies would be static structures and it would be impossible for organisms to exert control over slower environmental variables.

However, it is not broadly recognised that the birth, growth, death and renewal cycle, shown in Figure 2.2, transforms hierarchies from fixed static structures to dynamic adaptive entities whose levels are vulnerable to small disturbances at certain critical times in the cycle (Holling 1992a). The budworm example described in the previous section provides an example. That represents a transient but critically important bottom-up asymmetry that provides the opening for evolutionary change.

It has also not generally been recognised that, in addition, biogeochemical processes, as well as communities of organisms, adapt and "evolve" in the broadest sense of the word. Several examples of this have been shown for lakes, where organisms have been demonstrated to have significant long-term effects on chemical cycles. For example, Schindler (1977) showed that organisms are capable of increasing the nitrogen and carbon content of lakes fertilised with phosphorus to keep Carbon:Nitrogen:Phosphorus ratios in proportions suitable for algal growth. Sulphate-reducing and denitrifying bacteria are also capable of removing excess sulphate and nitrate, both strong acid anions, from lakes in order to maintain buffering capacities (Schindler 1980, 1986). In a sense, therefore, the organisms design a chemical environment suitable for life. It is Gaia (Lovelock 1988) at the level of an ecosystem.

To generalise, there are two key states where adaptive and evolutionary change is most likely to occur because slower and larger levels in ecosystems become briefly vulnerable to dramatic transformation because of small events and fast processes. One is when the system becomes overconnected as the ecosystem slowly moves toward maturity (Box 2, Figure 2.2). At these stages, there are tight competitive relations among the plant species. From an equilibrium perspective, the system is highly stable (i.e., fast return times in the face of small disturbances), but from a resilience perspective, *sensu* Holling (1973, 1987), the domain over which stabilising forces can operate becomes increasingly small. It is this combination that gives a brittleness to the ecosystem – it is strongly, but precariously controlled. It is tempting to see analogies in social systems i.e., of overcontrolled bureaucracies faced with crisis, of highly regulated businesses with terminal myopia, or of centralised empires near collapse at their peripheries.

Certainly at this stage in the ecosystem cycle, the system typically becomes an accident waiting to happen. In the boreal forest, for example, the accident might be a contagious fire initiated in a mature stand of jackpine, that becomes increasingly likely as the amount, extent and flammability of fuel accumulates. Or it could be a spreading insect outbreak, like the spruce budworm, triggered as increasing amounts of foliage both increase food and habitat for defoliating insects and decrease the efficiency of search by their vertebrate predators (Holling 1988).

Although human organisations have a host of ways to minimise, delay or eliminate similar large scale transformations, there are many examples of the potential for such transformations caused by increased brittleness – when for example, an environmental or resource government agency becomes more vulnerable to its critics from industry and environmentalists, or a business with declining profit margins and market share becomes more vulnerable to stockholder critics.

Small and fast variables can also dominate slow and large ones at the stage of reorganisation (Box 4, Figure 2.2). At this stage, the system is underconnected, with weak organisation and weak regulation. Here, instability comes because of loss of near equilibrium regulation rather than from the brittleness defined earlier for the shift induced from Box 2 to 3. As a consequence, it is the stage most affected by probabilistic events that allow a different entrained species, as well as exotic invaders, to become

established. An example is the invasion and extensive spread of an Australian tree *Melaleuca* into the Everglades of Florida after disturbance of
native vegetation.

Although Box 4 conditions represent the stage most vulnerable to erosion and to the loss of accumulated capital, it also is the stage which has
the potential to jump to unexpectedly different and more productive
systems. For example, when fire frequency declined along the prairie-
forest border in Minnesota during the Little Ice Age, 400 years ago
(Grimm 1983), extensive areas changed from oak savanna to maple
forest. It was the change in the disturbance or release function that
opened opportunity for rapid transformation of vegetation at the reorganisation phase.

Hence, at this stage, unexpected combinations of previously independent species can develop affinities among each other to give them a key
role in future structuring of the ecosystem. That is not too different from
the emergence of new innovative business when individuals with previously separate experiences combine their different skills to achieve a
new goal.

The degree to which small, fast events influence larger, slower ones is
critically dependent upon the accumulation, cycling and conservation of
accumulated capital. And that in turn depends upon the mesoscale disturbance processes. Human management of renewable resources or impacts
of macro-scale phenomena such as climate change, can release a pattern
of disturbance that destroys large amounts of accumulated renewal capital
over large areas. If too much capital is destroyed over too large an area,
the system can flip into a qualitatively different state that persists unless
there is explicit rehabilitation by management.

As an example, that is why grazing at sustained, extensive, but moderate
levels can transform productive savannas into less productive systems
dominated by woody shrubs (Walker et al. 1969). Or why successful efforts
of forest fire control can lead to so much accumulated fuel over such a
large area so that the inevitable runaway fire that eventually occurs, destroys accumulated soil and the capacities of trees to regenerate. That is
what led to the return of fire as a tool of management of the forests in the
U.S. national parks of the Sierra Nevada mountains. Previous successful
control of fire had led to conditions that threatened the maintenance and
reestablishment of such trees as the *Sequoia* (Holling 1980).

The question for issues of human transformation from the scale of
patches to the planet, therefore, is how much change does it take to release
disturbances whose intensity and extent are so great that the renewal capital is destroyed to the point where regeneration of plants (and thus habitats
for animals) are seriously compromised.

Roles for biodiversity

Any ecosystem – a forest, a lake, a grassland or a wetland – contains hundreds to thousands of species interacting amongst themselves and their physical and chemical environment. But not all those interactions have the same strength or the same direction. That is, although everything might be ultimately connected to everything else if the web of connections is followed far enough, the first order interactions that structure the system increasingly seem to be confined to a small number of biotic and abiotic variables whose interactions form the "template" (Southwood 1977) or niches that allow a great diversity of living things, to, in a sense, "go along for the ride." (Carpenter and Leavitt 1991; Cohen 1991; Holling 1992a). Those species are affected by the ecosystem but do not, in turn, notably affect the ecosystem, at least in ways that our relatively crude methods of measurement can detect. Hence at the extremes, species can be regarded either as "drivers" or as "passengers," although this distinction needs to be treated cautiously. The driver role of a species may only become apparent every now and then under particular conditions that trigger their key structuring function.

As a result, the impacts of humans or of climate often appear to be species specific where fundamental transformations are triggered from one biome to another – i.e. from forest to grassland, or grassland to desert, for example (Delcourt et al. 1983; Webb 1984). Geophysical processes have led to planetary changes in the past that were extreme enough to trigger profound shifts in climate and in vegetation. When those produced pronounced shifts between glacial and interglacial conditions, the vegetation was transformed and individual species interactions became uncoupled to form a variety of transient assemblages very different from either those that preceded the shift (Wright 1987) or those that now characterise major biomes (Davis 1981).

The important exception to this dissociated impact on individual species occurs when one or more of the critical structuring species and the associated physical conditions are affected so as to trigger fundamental transformations from one ecosystem type to another – i.e., from forest to grassland, or grassland to a shrubby semi desert, for example (Holling 1973; Walker et al. 1969).

In summary, we currently believe that only a small set of species and physical processes are critical in forming the structure and overall behaviour of terrestrial ecosystems, although there are insufficient long-term studies to deduce whether the presence or absence of rare species may cause slow, subtle shifts in ecosystem structure or function.

The fundamental point is that only a small set of structuring processes made up of biotic and physical processes are critical in forming the structure and overall behaviour of ecosystems, and that these form sets of relationships, each of which dominates over a definable range of scales in space and time. Each set includes several species of plants or animals, each species having similar but overlapping influence. It is the number of such species involved in the structuring set of processes that determines the functional diversity (Schindler 1990a, 1994). An example is the set of grass species that maintain the productivity and resilience of savannas (Walker et al. 1969), or the suite of 35 species of insectivorous birds that mediate budworm outbreak dynamics in the eastern boreal forest (Holling 1988).

Thus there is growing evidence that the great complexity and diversity within many ecosystems can be traced to a small number of critical structuring processes, some of which are mediated by critical "keystone" species (Paine 1966) or more commonly, sets of structuring species, and others by critical physical processes such as ocean currents (Roughgarden et al. 1991), or fire (Heinselman 1981), or drought (Walker 1981).

How could the great diversity within ecosystems possibly be traced to the function of a small number of variables? The many computer models of ecosystems or their parts that have been developed from process understanding and tested in nature, certainly generate highly complex behaviour in space and time from the interactions among a small number of variables, each of which operates at its own distinct speed (e.g., Holling 1986). For the models, at least, this structure organises the time and space behaviour of variables into a small number of cycles, presumably abstracted from a larger set that continue at smaller and larger scales than the range selected.

But are those features simply the consequence of the way modellers make decisions rather than the results of ecosystem organisation? This uneasy feeling that such conclusions can be a figment of the way we think, rather than of the way ecosystems function, is leading to a series of tests using field data to challenge the hypothesis that ecosystem dynamics are organised around the operation of a small number of nested cycles, each driven by a few dominant variables.

If there are, in fact, only a few structuring processes, their imprint should be expressed on most variables. That is, time series data for fires, seeding intensity, insect numbers, water flow – indeed any variable for which there are long term, yearly records – should show periodicities that cluster around a few dominant frequencies. In the case of the eastern maritime boreal forest of North America, for example, those periodicities were predicted to be 3–5 years, 10–15 years, 35–40 years and over 80 years (Holling 1992a). Similarly, there should be a few dominant spatial "footprint" sizes, each associated with one of the disturbance/renewal cycles in the nested set of such cycles. Finally, the animals living in specific landscapes should demonstrate the existence of this lumpy architecture by showing gaps in the distribution of their sizes and gaps in the scales at which decisions are made for location of region, foraging area, habitat, nests, protection and food.

All the evidence we have so far confirms just those hypotheses – for boreal forests, boreal region prairies, pelagic ecosystems (Holling 1992a) and for the Everglades of Florida (Gunderson 1992). Moreover, that evidence is in the process of being extended to demonstrate the existence of lumpy distribution of faunal body masses from specific sites in neotropical forests, from African savannas and in fossil assemblages occurring in North America just prior to the Pleistocene extinction of large mammals. A variety of alternative hypotheses based on developmental, historical or trophic arguments have been disproved in the fine traditions of Popperian science, leaving only the "world-is-lumpy" hypothesis as resisting disproof.

Therefore there is strong and growing evidence for the following conclusions:

- A small number of plant, animal and abiotic processes structure terrestrial biomes over scales from days and centimetres to millennia and thousands of kilometres. Individual plant and biogeochemical processes dominate at fine, fast scales; animal and abiotic processes of mesoscale disturbance dominate at intermediate scales (e.g., insect outbreaks, pulsed grazing by large herbivores, fire, wind); and geomorphological and global biogeochemical ones dominate at coarse, slow scales.
- These structuring processes produce a landscape that has lumpy (i.e., discontinuous) geometry and lumpy temporal frequencies or periodicities. That is, the physical architecture and the speed of variables are organised into distinct clusters, each of which is controlled by one small set of structuring processes. These processes organise behaviour as a nested hierarchy of cycles of slow production and growth alternating with fast disturbance and renewal (as is shown in Figure 2.4).
- Each cluster or "lump" is contained to a particular range of scales in space and time and has its own distinct architecture of object sizes, interobject distances and fractal dimension within that range.
- All of the many remaining variables, other than those involved in the structuring processes, become entrained by the critical struc-

turing variables, so that the great diversity of species in eco-systems can be traced to the function of a small set of variables and the niches they provide. The structuring processes are the ones that both form structure and are affected by that structure. *Therefore, these structuring variables are where the priority should be placed in investing to protect or enhance biodiversity.*

- The discontinuities that produce the lumpy structure of vegetated landscapes impose discontinuities on the behaviour and morphology of animals. For example, there are gaps in body mass distributions of resident species of animals that correlate with scale-dependent discontinuities in the geometry of vegetated landscapes. Thus these gaps, and the body mass clumps they define, become a way to develop a rapid bioassay of ecosystem structures and of human impacts on that structure. It therefore opens the way to develop a comparative ecology across scales that might provide the same power for generalisation that came when physiology became comparative rather than species specific.
- Conversely, changes in landscape structure at defined scale ranges caused by land use practice or by climate change will have predictable impacts on animal community structure (e.g., animals of some body masses can disappear if an ecosystem structure at a predictable scale range is changed). Therefore predicted (using models), planned (for land use development) or observed (using remote imagery) effects of changing climate or land use on vegetation can also be used to infer effects on the diversity of animal communities.

The lessons for both sustainable development and biodiversity are clear: focus on the structuring variables that control the lumpy geometry and lumpy time dynamics. They are the ones that set the stage upon which other variables play out their own dramas. That is, it is the physical and temporal infrastructure of biomes *at all scales* that sustains the theatre; given that, the actors will look after themselves!

That conclusion helps to define priorities for investment in protecting and enhancing biodiversity. But it should not be read as an argument that some species are important in nature but most are not. The species that presently appear to "go along for the ride" are the ones from which evolutionary processes can draw and develop alternative templates for the future. They may also be ones that humans value for their beauty, present or future medicinal properties, or other unique features. They may also play subtle, and as yet unforeseen roles in regulating ecosystem structure and function.

Like invention and innovation in an economy, there is an inherent unpredictability to evolutionary change. For example, had we been present in Jurassic times, it would have been impossible, *in principle,* to predict the eventual explosive ascendancy of mammals after the "age of the dinosaurs." It is only possible to explain the phenomenon after the fact. The diversity that the small set of structuring processes and species foster, provide the options for such unpredictable futures. But focusing exclusively on those species in attempts to preserve biodiversity is simply a recipe for diluting time, energy and money. The target and the priority should be on the ecosystem structure at specific scale ranges and on the processes that form that structure.

2.5   Marine vs. terrestrial ecosystems

The synthesis we describe to this point is largely one appropriate for terrestrial systems or for aquatic ones dominated by substrate i.e., near-shore ecosystems with abundant recruitment, coral reefs or shallow fresh water lakes. Open pelagic marine systems are different, because of the different physical properties of land and air on the one hand, and water on the other.

The planet's biota has two strategies for dealing with variability. One is to adapt to the existing variability in time and space. The other is to control that variability, and by so doing, introduce states that are able to absorb specific ranges of variability without flipping into a different state. Terrestrial ecosystems do demonstrate a measure of such control of variability over some, but not all scales, that gives them an inherent resilience. Lakes do as well, as exemplified by their ability to modify nutrient ratios and buffering capacities, as discussed above. It is useful to identify the significance of that control by contrasting terrestrial ecosystems with their opposite, i.e., pelagic ecosystems of the oceans.

A hierarchy of structures, similar to that in forests, occurs in the oceans. These structures are produced by physical forces and result in a nested set of levels that increase in scale from turbulence, to waves, to eddies, to gyres. As in a forest, animals respond by adapting to the structure of those levels – for example, phytoplankton to fine scales, predatory fish to coarse. The big difference between oceans and terrestrial systems, however, is that the biota directly control the structure of terrestrial ecosystems over a definable range of short temporal scales, whereas the pelagic organisms of the ocean are stuck with the necessity of adapting to existing physically determined variability in the short as well as the long term (Steele 1985, 1989). The physical properties of water and absence of fixed substrate leave no alternative except for pelagic organisms to adapt to variability imposed by physical processes.

As an example, the life cycles of the organisms in coastal ecosystems pass through a pelagic oceanic phase where the controlling ecological processes are physical rather than biotic. The importance of this pelagic phase in determining inshore patterns of diversity has only recently been recognised, and has discredited the earlier view of coastal ecosystems being closed systems whose patterns were determined largely by food web relationships. Today, marine ecology is as much the physical oceanography of currents and winds as it is the biology of eating and mating.

Terrestrial community ecology is also witnessing a renewed recognition of physical processes as controlling ecological community structure, especially with regard to slow processes having time scales of 10^4 to 10^7 years, such as the sea level changes and plate tectonics usually studied in geology. Although ecology and evolution were closely associated with geology in the first fifty years of this century, since the 1950s until quite recently ecology has been largely viewed as one of the levels of organisation in biology. It was assumed that population dynamics and species interactions are fast enough to erase the initial condition established by geology, as though the ecological play were independent of the contractor who constructed the theatre. In fact, the ecological play is taking place as the theatre is being built, with the actors adjusting their roles to the state of construction, and occasionally joining the carpentry. As a result, some community ecologists argue that ecology is not correctly viewed as a part of biology at all, but is instead an organic earth science.

If terrestrial biota can exert control over fast variability up to perhaps a few decades but cannot control very slow geophysical processes, and marine communities are controlled typically only by physical processes at all scales, what form shall generalisations in ecology take, including generalisations across terrestrial and marine ecosystems? In biology, generalisations rely on a common mechanism, as exemplified by the convenience of DNA being the chemical basis of heredity in most living things. In the elementary physical sciences, generalisations are theoretical laws, with empirical cases being instances of general laws. But what of the earth sciences, including ecology? There is no one mechanism that makes volcanoes, nor is there an ideal general river of which all real rivers are imperfect instances. Instead, generalisation in the earth sciences consists of finding a small collection of big particulars. There are, after all, only seven continents, four large ocean gyres, two hemispheres, and one Earth. In the earth sciences, the finiteness of the earth itself limits the number of particulars, provided the particulars are sufficiently big. Thus ecologists can offer consensus on how a small collection of ecosystems work, even though such statements do not take the form of universal generalisations.

To illustrate, we describe how an oceanographic mechanism controls the dynamics of coastal populations along California. This coast is comparable to the world's three other major up welling systems off Peru, Portugal and Northwest Africa, and Southwest Africa.

Ecology of the intertidal zone

Experiments undertaken during the 1970s in Scotland and the Pacific Northwest of North America yielded three generalisations concerning rocky intertidal communities. First, the organisms that compete for space on the rocks can be ranked in a linear hierarchy of competitive ability. For example, a mussel can overgrow and smother a medium-sized barnacle, which, in turn, can overgrow and smother a smaller barnacle resulting in a tendency for the system to culminate in a monoculture of the dominant competitor.

Second, disturbance (here considered to be any abiotic source of mortality to organisms that occupy space), if not too high, prevents the dominant competitor from excluding the subdominants, thereby allowing diversity to persist. If the disturbance is high enough though, the habitat becomes unlivable. Therefore, an intermediate degree of disturbance is best for maintaining diversity. This proposition is known as the "intermediate disturbance principle."

Third, a predator, which frequently prefers to consume the dominant prey, can perform much the same role as disturbance in ensuring that diversity exists in the system. If the predator is removed, the dominant prey can proceed to exclude its competitors, resulting in a monoculture. The predator is therefore known as a "keystone species" because its removal causes a loss of species diversity in the system (Paine 1966).

It is now appreciated though, that these generalisations were obtained from a special class of sites that happen to enjoy a high rate of larval settlement. They function much as does a terrestrial ecosystem, dominated by substrate conditions and biotic interactions. The full ecosystem, however, consists not only of the adults that compete for space and incur predation and disturbance as just discussed, but, as well, the larvae that live in the ocean adjacent to the coast. These larvae feed while in the ocean, and may be carried scores of miles offshore before being returned by currents to the habitats where they can metamorphose into adults. Therefore, the processes affecting larvae in the ocean could be every bit as important as the processes affecting the adults on the shore. Indeed, if the rate of larval arrival of the dominant competitor to the shore is infrequent enough, the dominant competitor never has a chance to attain sufficient abundance to exclude the subdominants. With limited larval arrival, the dominance hierarchy becomes moot, and both the intermediate disturbance principle and keystone predator effect cease to function. The error of the early work was in over generalising to all marine communities from sites that happened to be physically situated to enjoy large larval return rates.

The study of the mechanisms controlling larval return rates is now assuming great interest because less is known about this part of the ecosystem than about the adults so easily observed during a stroll among the tide pools on a weekend visit to the shore. The mechanisms controlling the larval return of coastal marine populations have a dominant characteristic – their propensity for large fluctuations in abundance. An early model of theoretical ecology was proposed by Vito Volterra in 1931 to explain the fluctuations he witnessed in the fish markets of Trieste. It later turned out, using time series analysis, that the fluctuations were not regular enough to be viewed as an oscillation as postulated by Volterra, and the cause of marine population fluctuations has remained mysterious.

Along the California coast the cause of population fluctuation lies somewhere in the ocean and not on the coast itself. With intertidal barnacles for example, adults attached to the rocks steadily release larvae into the sea during the spring and summer. Yet the mature larvae arrive back to the intertidal zone in episodic pulses that bring about large fluctuations in abundance. Determining the cause of episodic recruitment of larvae into the adult stock has come to be known as the "recruitment problem" in marine ecology. Solving this problem involves two subproblems: finding the mechanism that concentrates the larvae into discrete aggregates or "patches," and finding the mechanism that determines when the patches of larvae arrive at the shore causing a recruitment pulse.

Studies in recent years have tested a specific hypothesis that appears to solve the recruitment problem for intertidal barnacle populations along the California coast (Roughgarden et al. 1991). As the cold and saline water upwells adjacent to the coast and moves away toward the open ocean, it intersects the California Current, forming a frontal boundary. This frontal boundary is a convergence zone and it is hypothesised that planktonic organisms, including barnacle larvae, accumulate there. The location of the front moves according to the winds that control the strength of upwelling. When the winds are strong, the front is pushed far from shore and transport of upwelled water at the surface carries intertidal larvae away from shore. As the winds weaken, the strength of cross-shelf transport weakens and the front moves closer to shore. If wind relaxation occurs for an extended time the front apparently can collide with the coast and deposit its accumulated larvae producing a recruitment event. This hypothesis potentially explains both the timing of recruitment at the coast and why it occurs in discrete pulses.

Evidence for this hypothesis has now come from three sources. First, the ocean off California contains larvae of many species of barnacles, and of these several types come from adults that live in the intertidal zone along the coast, and several other types come from adults that live attached to driftwood and other surfaces in the California Current. Plankton samples from cruises showed the existence of a front separating the offshore pelagic barnacle larvae from the inshore intertidal barnacle larvae. Moreover, this front moved away from shore when the winds were strong and back toward shore when the winds were weak (Roughgarden et al. 1988). Second, recruitment pulses along the California coast were observed to coincide with the arrival at the shore of the warm low-salinity water typical of the California Current – this replaced the cold high-salinity water normally found adjacent to the coast (Farrell et al. 1991). Third, direct data that larvae are concentrated at an upwelling front and that the collision of a front with the intertidal zone caused a pulse of larval settlement were obtained with field studies (Roughgarden et al. 1994).

Thus, the case is building for a synthetic picture of how the dynamics of larval rocky intertidal barnacles are controlled in central California. The larvae are released throughout the spring and early summer. They are carried offshore to fronts separating the upwelled water from the warmer and fresher water of the California Current. There the larvae feed on the abundant algae of the nutrient-rich upwelled water. Then, as the upwelling season wanes, the fronts migrate to shore returning the larvae to the adult habitat where they settle and metamorphose into small adults. The strategy is similar to freshwater systems in which the larval stage is placed in a transient habitat with high food (and perhaps also high predation risk) such as salamander and frog tadpoles in temporary ponds and streams.

The type of model that can be used to predict the population dynamics of a marine population consists of simultaneous equations for both the larval and adult phases (Roughgarden et al. 1994). The processes affecting the adults include the familiar ecological interactions of competition for space, predation and disturbance. The processes affecting the larval phase include the currents that cause larvae to be aggregated at fronts, and the processes that move the position of fronts. The form of the model consists of an ordinary differential equation(s) for the interactions among the adults in the benthic site, and a diffusion/advection equation(s) for the transport of the larvae in the ocean. The offshore upwelling front can be modelled as a reflecting boundary. The physical parameters in the model are thus the location of the reflecting boundary indicating the front, and the magnitude of the advection in the surface waters. Data on the position of offshore fronts and on the vector field of advection in the surface mixed layer of the ocean can be obtained from remote sensing technology. The AVHRR sensor in the U.S. NOAA satellites yields images of sea surface temperature, and fronts are located in the images as locations of high gradients in surface temperature. The technology yields a map of current vectors for a semicircle with a radius of about 50 km from the station. These data can be assimilated every 3 h, say, and used to forecast the time series of recruitment and population dynamics of coastal populations.

Generalisations

As for generalisation across marine and terrestrial systems, what can be ventured? Both marine and terrestrial communities are increasingly viewed as open systems. Sometimes the organisms transport themselves across the system boundaries, as if acting as their own agents. More often, the transport is brought about by, or permitted by, physical processes taking place in the habitat or its vicinity.

In marine ecosystems the transports involve currents, and more specifically, the motion of fronts at which organisms have been concentrated. These transport mechanisms enter in models at the population scale, and then are propagated to the community scale and higher.

Marine ecosystems are more open than terrestrial systems because water provides both the substrate and dispersal medium, whereas, for terrestrial systems, air is visited only for brief periods. Hence, by increasing the perimeter of a terrestrial site's boundary, the site becomes increasingly closed, whereas marine sites, quite large by terrestrial standards, are still open systems because of the longevity of the dispersive phase. As a result, an ordinary differential equation of a kind familiar in the traditional ecological literature may be satisfactory for a terrestrial population, provided the site is taken to be sufficiently large. (A caveat is that if the site is taken to be too large, it is internally heterogeneous, requiring treatment as a metapopulation of interconnected sites.) In contrast, at the community scale, transport mechanisms such as wind for spores and bacteria, land bridges for mammals, and plate tectonics for reptiles must be explicitly considered for community patterns. Ecological interactions simply do not explain much about community structure unless the cast of characters is known, and the casting is carried out by these transport processes.

These generalisations suggest six policy implications. First, terrestrial systems are functionally more localised than marine systems, and extinctions more likely to result from habitat development. Second, a widespread terrestrial biome such as a forest is not functioning as a large single system but as many instances of the same system that happen to be contiguous. In contrast, the similarly widespread rocky intertidal zone is functioning more as a single ecosystem because of the large spatial scale of the larval transport across it. Third, damage to a marine ecosystem is likely to be discerned farther away from the spot of the stress than in a terrestrial ecosystem. Logging in streams affects fishing stocks hundreds of miles away where different industries and constituencies are involved. Damage to a terrestrial ecosystem is more likely to be discerned near where the damage occurred – soil runoff impacts the very neighbourhoods where the clear cutting has taken place. Fourth, recovery of marine ecosystems is faster and more possible when stress is removed because recolonisation of the stressed area is faster and components are less likely to be lost permanently. In a terrestrial ecosystem full recovery is impossible if extinctions have occurred. Fifth, precious ancient life forms and species combinations are more common and more endangered on land than at sea. Sixth, management of renewable resources on land or at sea requires adaptive strategies of management in order to monitor both changing external variability as well as unpredictable internal changes induced by management transformations.

2.6   How human impacts change biodiversity

Before delving further into the effects humans have on biodiversity, we need to establish what we mean by the term. It has been much abused and has become almost meaningless in its political usage. Scientifically, however, it is generally accepted to be some product of three levels or scales of diversity – genetic, species and ecosystem (or community). The ecosystem or community level has to do with the spatial scale and pattern of species combinations (their patchiness), and the species and genetic levels encompass the numbers of species and the variation within them. Defined in this way, biodiversity assumes that all species and genotypes are of equal significance (i.e., they contribute equally to the value, or measure, of biodiversity) though some analyses take phylogenetic distinctness into account (e.g., Vane-Wright et al. 1991), giving more weight to species that are very different to the others in a community, in terms of their evolutionary history, or taxonomic distance.

But biodiversity is more than this. Especially when considering the impacts of humans, biodiversity concerns much more than just the dramatic loss of species. Another way of looking at the diversity of life in ecosystems, is to consider it in terms of the diverse functions that are performed. Such functional diversity is more difficult to identify and classify, since the differences are not easily observed and may only become apparent at particular times. Nevertheless, the difficulties involved in a functional classification do not deny its importance, and changes in functional diversity will have profound and immediate effects on ecosystem performance. We described in the previous section how loss of a keystone species (by definition, a functionally distinct biological entity) will have a much greater effect on other species and on ecosystem function than will the loss of a species whose ecosystem role is also performed by several others. To someone interested in the uniqueness of species, in their own right, the two kinds of species have equal inherent value. But if the interest extends to how to maximise the likelihood of all species persisting, then we need to minimise the likelihood of changes in ecosystem function, and in this regard those functional groups which are represented by just a single species warrant special conservation attention (Schindler 1990b).

Because function is more difficult to identify and measure, and does not easily lead to an ordered, hierarchical classification, ecologists have so far been unsuccessful in their attempts to develop an operational functional classification of plants and animals. It is currently a topic of great interest, particularly in regard to predicting the effects of climate change. However, in the absence of a workable and generally accepted classification of plant and animal functional types we will proceed by considering the effects on biological diversity in terms of the direct and indirect effects on the three levels described earlier – genetic, species and ecosystem – and consider these in terms of their functional significance.

Genes

Human activities leading to a decline in intraspecific genetic diversity may be a serious problem under particular circumstances, but overall it is the least serious of the three levels of effects. Considering all the issues underpinning conservation biology and the maintenance of biological diversity, we concur with the view that while it may be important in some cases, the decline in genetic diversity within species is of relatively low priority compared to the problem of declining populations and the widespread local loss of species and (we would add) habitats.

Species

Loss of species, both locally and in totality, is a major concern. Human activities have accelerated species loss in a variety of ways, and the record confirms that recent extinctions (i.e., the last couple of hundred years) are well in excess of those that would be expected under evolutionary time scales in the absence of a catastrophic global environmental effect. The factors that have had the greatest impact on species loss include loss of habitat (conversion to "agroscapes" or urban complexes), harvesting/ hunting, altered fire regimes, direct changes in herbivory (e.g., by livestock grazing), introduced predators and diseases, home-range/habitat fragmentation and atmospheric changes (e.g., impacts on species in soft water lakes because of acid rain). As described earlier in the sections on temporal and spatial effects, these factors seldom act alone or in a uniform, steadily progressive (or regressive) way. The losses or declines occur in particular places and under particular conditions. In arid regions (central Australia, for example) resource-rich refuge for animals occur in a background mosaic of resource-poor areas (where resource = water and nutrients). The dependence of animals on these richer patches is most pronounced during characteristic widespread droughts (Stafford-Smith and Morton 1990), and grazing by domestic livestock at these times, in these patches, has led to local extinction of many species and contributed to the total extinction of some.

Attributing blame for an observed decline in a species to any one of the above factors is made difficult by the fact that they seldom act alone. Humans influence a region in a variety of ways, usually bringing about simultaneous changes in landscape patterns, fire regimes and harvesting/ hunting/grazing regimes, and for good measure generally introduce a range of exotic species. The loss of the burrowing bettongs (*Bettongia lesueur*) in Australia (a rabbit-sized and rabbitlike marsupial) has been attributed by various conservationists to competitive displacement by rabbits, to fox predation and to altered habitat through changed fire regimes. All three are probably involved. In a similar vein, the semiarid savannas of the world have exhibited a general trend of increased woodiness under grazing, often to an extent where the change is irreversible without massive intervention, and two mechanisms are involved. First, grazing reduces uptake of water from the topsoil by the grass layer (G), reducing its suppressive effect on woody seedlings and allowing successful establishment and subsequent rapid growth of woody plants (W); second, fire is prevented, either as a management decision (because fire "competes" with the livestock for grass fuel), or because grazing levels are such that insufficient fuel accumulates to carry an effective fire. The interactive effects of the two are as follows.

The long-term, equilibrium mix of G and W in savannas depends on soil type. In sandy soils it is inevitably a woodland, since W has a competitive advantage over G in all phases of growth, and only fire maintains an open state. On heavy soils, the grass layer is dominant over woody seedlings and, provided there is always a full, vigorous cover, W is prevented from establishing. Once established, however, W dominates G, but because the response time of W is slow relative to the large interannual variations in water supply (it takes years for a tree to grow compared to the very fast response of the grass layer) sustainable biomass of trees is lower than the average rainfall could support. There is therefore usually sufficient available topsoil water for a grass layer to be maintained – further reducing the water available to trees. Provided G is not reduced too much by grazing, periodic fires reduce W more, allowing a larger $G:W$ ratio. The two factors, fire and water, are strongly interactive in their effects (Scholes and Walker 1993), and human actions in regard to livestock grazing and control of fire lead to a series of interactive effects which alter both the structure ($G:W$ ratio) and species composition of the savanna.

Finally, growing modification of the atmosphere because of pollutants, represents a distinct class of influence that transfers local effects into aggregate subcontinental effects. An example is provided by studies of softwater lakes affected by acid rain. Schindler et al. (1991) and Minns et al. (1990) have shown that in lakes of the most acidified regions of eastern North America, the number of species lost due to lake acidification may be several tens of percent in a given ecosystem. When multiplied by the number of lakes (over 600,000 in areas of eastern Canada affected by acid rain alone) the numbers are impressive. For example, over 200,000 populations of fish and one million populations of invertebrates are estimated to have been lost due to acidification (a population is defined as one species in one lake).

Ecosystems

The most pronounced direct effects of human activities have been on ecosystem diversity, largely as a consequence of land clearing. In all of the world's biomes, humans have selectively cleared the most fertile areas for conversion to agriculture, resulting in very significant losses (in some cases complete loss) of valley bottom and lower catena systems – the zones of nutrient and water accumulation. The lowland tropical rainforests, the high fertile areas in temperate forests (cf Brathwaite and Walker 1989) and the deeper soil, low-catena positions in the savannas have been targeted for agriculture. In addition to direct loss of particular biological communities this process has also led to fragmentation of the remaining areas and to changes in the spatial scale and pattern of the landscape. The importance of this for component species was described earlier.

Indirect effects

Although many of the effects described above result directly from human activities, there are many more which come about as a result of secondary effects. Indirectly, human activities influence both species and ecosystems in two main ways. First, via changes in the environment – in atmospheric composition (acid rain, CO_2 increases, climate change), changes in hydrology (altered run off, soil moisture, river flow regimes) and changes in nutrient inputs (fertiliser drift, nitrogen accumulation). For example, acid deposition and increased nitrogen levels (via rainfall) in European forests result in a complex sequence of changes in soil chemistry and tree growth response, largely dependent on changes in magnesium availability and in the form of soil nitrogen. This has led to reduced growth and dieback of trees, and enhanced growth of nitrophilous species in the understorey, altering species composition (Schulze and Ulbrecht 1991). In some areas it has resulted in such vigorous growth of grasses on the forest floor that it has effectively stopped tree seedling establishment (Schulze personal comment), and is therefore preventing forest regeneration. This provides yet another example of a multistable system and of a flip into an irreversible state with loss of resilience.

Similarly, wind erosion in the western wheatbelt of Australia (rather infertile, old soils) leads to fertiliser (phosphate) drift into remnant patches of native vegetation and shifts the competition balance between native species and introduced weeds in favour of the latter (Hobbs and Atkins 1988).

Another suite of indirect human-induced changes in diversity have been brought about through changes in biological interactions, via the impacts of introduced species. Four main processes are involved:

- *Competition* – the many examples of plants, insects and vertebrates that have competitively displaced native species, very often because they arrive in their new habitats without the suite of pathogens and herbivores that regulate their numbers in their own, native habitats.
- *Predation* – for example, the introduction of snakes to Hawaii, and (as already mentioned) foxes to Australia, in both cases leading to extinction of some animal species and to considerable local changes in the relative abundances of many others.
- *Selective mortality through disease* – one example is the root fungus (*Phytopthera cinnamomi*) that has caused significant changes in plant species composition in Australia (Wills 1993).
- *Environmental change* – for example the effects of the introduced

nitrogen-fixing legume, *Myrica faya,* in Hawaii, which has altered the competitive balance amongst native species (Vitousek and Walker 1989).

Unseen changes and functional consequences

Within the range of effects just described, the changes in biodiversity may frequently go unnoticed, owing to superficial adjustments in the biological community. Subsequent changes in the performance or composition of the community then come as a surprise to those who, perhaps unwittingly, caused them. As an illustration, in the semiarid rangelands of the world a major problem confronting livestock owners is the large interannual variation in fodder production in response to variation in rainfall. It varies by five-fold or more from one year to the next (Barnes 1979) and, as a result of changes in biological diversity brought about via their management of livestock, humans have frequently brought about an increase in this variation. The most common way this has happened has been through loss of perennial grasses and their replacement by annuals, which vary far more in response to fluctuations in rainfall than do perennials (Kelly and Walker 1976). Of comparable significance, though, is the disruption of a somewhat more subtle ecological mechanism that counteracts inter-annual variation in production, i.e., a high phenological diversity in the grass sward. If there are approximately even amounts of early, mid- and late season growing grasses then, whenever the rains fall, there will be about the same amount of grass that is best able to respond to it (Walker 1988). Heavy grazing pressure, however, can lead to the loss of early season palatable species with their place being taken up by later growing species (Silva 1987). Where this occurs, although there may still be about the same amount of grass cover, the loss in phenological diversity means that in years of mostly early season rains production will be less than expected, and vice versa for years of mostly late rains.

A significant feature of this change in rangeland diversity and function is that the grass sward still appears to be healthy, and because replacement of the early growers by the later growers takes place slowly, and often in fits and starts, the change may go unnoticed to anyone who is not closely monitoring the trend (as opposed to the fluctuations) in species composition. The average biomass is much the same, but because diversity (in this case functional and species diversity) has been reduced, the inter-annual variation in fodder production can be greatly exacerbated.

2.7 Conclusions

The common view of human impact on biodiversity is one of species loss. While not diminishing the importance of this in any way, or indeed the importance of genetic diversity loss, by far the greatest impact is the complex set of changes in the proportional species composition of communities, particularly those involved in critical structuring processes. The last example illustrates some of the complexity involved in the relations between ecosystem function and biological composition. We stated in the introduction to this chapter that an ecosystem under stress apparently keeps its function even though species composition changes. We need to add some qualifications to this. First, there is obviously a limit to the amount of species change that can be sustained before function changes, and the kinds of species involved in the change determine this limit. Second, we need to distinguish between complete or absolute change in the species that are present, and changes in their relative abundances. In keeping with the last example, McNaughton (1985) has shown that in the Serengeti grasslands those communities that varied least in year-to-year production showed relatively greater changes in species contribution to the biomass. In other words, those communities that were able to exhibit differential species responses to a fluctuating environment were most constant in terms of primary production. The same suite of species were present all the time, but different species responded and contributed most to the biomass in different years. Finally, the interactive effects of structure and function are clearly neither linear or additive and in terms of human effects on biodiversity and its consequences for ecosystem function, it is not a loss in species or a change in some diversity index, per se, that matters, but rather the kinds of species that are affected and their role in structuring ecosystems and maintaining resilience.

CHAPTER 3

Scale and biodiversity in coastal and estuarine ecosystems

Robert Costanza, Michael Kemp and Walter Boynton[1]

3.1   Introduction and summary

This chapter reviews coastal and estuarine ecosystems (with a special emphasis on the Chesapeake Bay) in terms of their unique biodiversity characteristics. Coastal and estuarine systems are generally low in species diversity. The exceptions are coral reefs and reefs that form on artificial structures. We attribute this to the general unpredictability of the estuarine environment. Reefs generally form in more predictable environments. An unpredictable environment selects for generalist species in order to cope with the unpredictable changes. In unpredictable terrestrial environments, ecosystems can build structure (i.e., trees and soil structure) to smooth out some of the noise and create a more stable base for increasing biodiversity. But estuaries are so dominated by the physical forces of water flow that these structures are impossible to build. Estuaries are characterised by their relatively small standing ecological structure and the high degree of organism mobility. All of these characteristics point to a high degree of ecosystem "resilience" (Holling 1986), and in fact we observe this high resilience in most estuaries.

We give a detailed description of the Chesapeake Bay and its watershed to help define these ideas and extrapolate them to the task of managing complex ecological economic systems. Biological or species diversity is put in a systems context as a scale-dependent measure of an important system characteristic, but one that may be inappropriate as a management focus in estuaries because of their special characteristics. In estuaries it is the diversity of ecological *processes,* and in particular certain keystone processes, that are more critical and that should be the focus of management efforts.

[1] Lisa Wainger and John Woodwell provided considerable editorial assistance and comment on drafts of this chapter. Carl Folke and Joy Bartholomew provided useful comments and suggestions. The Beijer International Institute for Ecological Economics provided cheerful support during the preparation of this manuscript.

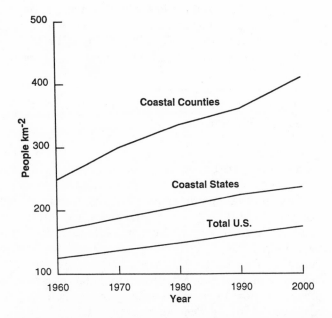

Figure 3.1. Population density of coastal counties in the United States compared with coastal states and the total United States from 1960 to 2000 (projected). Data are from LMER Coordination Committee (1992).

We conclude with a hypothesis that the diversity in a system is a function of the predictability of the resource environment on the scales at and below the scale of the system of interest, as mediated by the evolutionary developmental dynamics of complex systems. We believe that this hypothesis is consistent with the data on diversity in estuaries and other systems, and can be tested further in the future via comparative analysis.

3.2   Coastal and estuarine ecosystems

Coastal and estuarine ecosystems are the vast biomes which join continental lands and oceanic islands with their surrounding seas. Most of the world's population resides in the coastal zone, and as Figure 3.1 indicates, the density of coastal economic development is increasing. Therefore, these ecosystems are particularly important for integrating sound ecological management with sustainable economics.

Estuarine ecosystems are coastal indentations with "restricted connection to the ocean and remain open at least intermittently" (Day et al. 1989). Salinities are usually intermediate between those of fresh and seawater, but in regions where evaporation is high or rainfall low, estuarine salinities may be equal to or higher than those of the ocean. Most present day estuaries were formed during the last 15,000 years of the current interglacial period, and are thus geologically ephemeral features of the landscape (Day et al. 1989).

Although estuaries vary in depth from less than one metre to several hundred metres, their shallowness clearly distinguishes them from the open ocean. Depending on their origin and the nature of their surrounding land masses, estuaries may assume a variety of sizes and forms. Many (e.g., coastal plain estuaries) were formed as a result of the drowning of coastal rivers by rising sea level, while others (e.g., fjords) were formed in glacial channels with terminal sills associated with moraine deposits. Other estuaries occupy the chasms left from tectonic activities or were formed as part of river deltas. Estuaries also include the large system of shallow coastal lagoons formed from oceanic sedimentological processes behind barrier islands, peninsulas and spits. The nature of these coastal ecosystems depends largely on whether they are situated on the continental shelves of western or eastern oceanic boundaries or on peripheries of oceanic islands. In some cases, like coral reefs, the ecosystems themselves create the estuarine environment.

Estuaries differ substantially from the ecosystems occurring at their oceanic and freshwater boundaries. In many temperate regions, estuaries are wetland ecosystems bordered by emergent grasses and shrubs, while in tropical environments these wetland edges are often dominated by halophytic (salt tolerant) trees, called mangroves.

One of the largest and best studied estuaries in the world is the Chesapeake Bay. Because of the special characteristics of the Chesapeake Bay and its watershed, it represents a good case study for further elaboration of some key concepts in estuarine science and management, and in particular issues of scale and diversity. We will use the Chesapeake Bay to supply detailed examples of issues in the discussion that follows.

Special physical characteristics of coastal and estuarine systems

Perhaps the feature that most distinguishes estuaries from all other ecosystems is the nature and variability of the physical forces which influence them. Within small geographic regions, many estuaries experience widely varying and unpredictable conditions of temperature, salinity, and water movement over relatively short time scales. Although fluctuations in some physical features, such as temperature, are damped in marine systems compared to terrestrial environments (because of the large heat capacity of the water mass), the shallowness of estuaries makes them more susceptible to larger amplitude variations. In estuaries, local short-term variations of salinity and water mixing are substantial.

One can get a feel for the nature of these variations by plotting the frequency of the variations against their intensity, as shown in Figure 3.2. In both marine and estuarine environments, the intensity of variations in many physical forces exhibit patterns which are inversely related to the frequency (Figure 3.2b). This so-called red noise distribution means that higher intensity events occur less frequently and there is a gradual decline in intensity with increasing frequency. This pattern is associated with systems that have a high degree of interaction between events at low and high frequencies and it characterises both the full range of estuarine system variations (Figure 3.2b) and terrestrial processes at scales longer than 50 years (Figure 3.2a, Steele 1985). For higher frequency events in terrestrial systems (which occur within the lifetime of many organisms), variations in physical forces tend to be uncorrelated with the frequency of occurrence, producing a pattern of "white noise" (Figure 3.2a, $f > 10^{-2}$). One ecological consequence is that organisms in terrestrial environments (as opposed to estuarine) are better able to adapt their behaviours and physiologies to the more predictable physical variations in their habitat.

Ecological diversity of estuaries

The relatively large and high frequency variations in salinity and water movement characterising most estuaries tend to limit the number of animal and plant species capable of adapting to these rigorous conditions (Day et al. 1989). As an "ecotone" between fresh and marine environments, estuaries contain a mixture of freshwater and oceanic species, but both planktonic and benthic communities contain substantially fewer species than do similar communities in oligotrophic lakes and in the ocean. The number of benthic faunal species in the deep sea may be comparable to that of tropical forest biomes (Grassle 1991), while estuarine benthos are commonly dominated by a few species, with limited taxonomic diversity in any given bottom area (e.g., Boesch 1974).

The relatively low taxonomic diversity of estuarine communities arises because of physiological difficulties in dealing with high-amplitude unpredictable stresses (Slobodkin and Sanders 1969) and because of the high organism mobility afforded by physical processes and properties of water. Osmotic stress, or the need to maintain balanced levels of water and salt internally, is a primary physiological limitation for many organisms. In an estuary, organisms must adapt to continually changing salinity. The cost of the morphological apparatus to handle these changes takes a big share of the estuarine organism's available potential energy and limits the num-

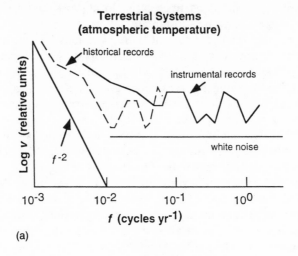

(a)

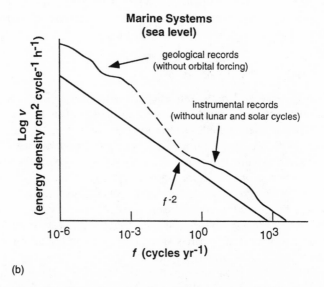

(b)

Figure 3.2. Comparison of variance frequency spectra for (a) terrestrial systems and (b) marine systems. Data are from Steele (1991).

ber of species capable of surviving in this environment. Other factors such as oxygen deficiency can also contribute to environmental difficulties. Estuaries do provide a means for rapid dispersion and easy mobility for resident organisms, due to the buoyancy of water and the rapid hydrodynamic transport associated with tides, winds and pressure gradients. Many coastal marine organisms have high fecundity and depend on water transport for larval dispersal, so that isolated populations are rare. Many fish and swimming invertebrates migrate long distances (10^2–10^3 km) during portions of their life-cycles. Consequently, functional replacements for a given estuarine species are virtually always available (Steele 1985), so that the selective advantages of specialisation are minimal. There are relatively few endemic species in estuaries; most originate from freshwater or oceanic environments and many marine organisms require estuarine environments for a portion of their life-cycle.

In addition to the physical stresses, another factor which limits the number of species in coastal marine environments is the virtual absence of the physical structures (and associated habitats) created by organisms (e.g., plant canopies) which typify terrestrial environments. In very shallow coastal environments (generally less than 10 m), however, rooted vascular plants (sea grasses) and attached algae (e.g., kelp) often do create complex physical structures which lead to multiple physical niches and relatively high taxonomic diversity. Similarly, reefs formed by colonial animals (oysters, mussels) in estuaries can produce complex physical structures containing relatively more species. In very stable tropical marine environments, coral reefs develop unparalleled taxonomic diversity. Coral reefs cannot survive in highly variable environments, and such high-diversity systems do not occur in estuaries.

In general, the relationship in any ecosystem between taxonomic diversity (number and distribution of species) and functional diversity (variety of ecological processes) is unclear. The basic ecological processes involved in biogeochemical cycles and trophic interactions are the same in coastal marine ecosystems as in any other biome. In estuaries, however, a given species is relatively less specialised for performing a single (or a limited repertoire of) ecological function(s). For example, the cosmopolitan estuarine clam, *Macoma balthica,* acts as a suspension feeder (filtering food from the overlying water) in environments with low rates of organic deposition to sediments, but acts as a deposit feeder (scavenging food from the sediments) in organic-rich environments. Indeed, most estuarine animals appear to be opportunistic feeders, altering their diets to focus on foods which are relatively abundant (Darnell 1961). Similarly, most estuarine bacteria have alternative metabolic pathways for obtaining energy (Fenchel and Blackburn 1979), and the same algal species can be found dominating benthic diatom, phytoplankton or epiphytic communities under different estuarine conditions (Day et al. 1989). One measure of the functional diversity of marine ecosystems is the variety of different responses displayed by their organisms to the range of physical environmental changes that they confront (Steele 1991). Thus, the relatively large scales of physical (and attendant biological) variability in estuaries (Powell 1989) might suggest that, despite their low taxonomic diversity, estuarine ecosystems have high functional diversity.

Estuarine productivity

Estuaries and coastal marine ecosystems are cited among the most productive biomes of the world (Mann 1982). One reason for the high primary productivity of estuaries is the relatively high nutrient loading rates compared to agricultural systems and other biomes (Figure 3.3). Rates of carbon fixation in coastal marine ecosystems rival those reported for the most productive terrestrial environments and substantially exceed those for oceans and many lakes (Figure 3.4). While benthic algal and sears can contribute substantially to estuarine production in shallow and clear coastal environments, phytoplanktonic algae tend to be the dominant autotrophic group. When one considers primary productivity in terms of nitrogen assimilation, however, estuaries and eutrophic lakes are by far the most productive environments on earth (Kelly and Levin 1986), because only 10% as much nitrogen is needed to produce algal tissue as is needed for tissue of trees and woody shrubs. In addition to the relatively high rates of nutrient inputs to estuaries, their shallow depth (proximity of sediments to euphotic zone) promotes efficient nutrient recycling. Physical circulation, characterised by landward flow of flooding tides and particle trapping with density-driven stratification, leads to efficient nutrient retention (Kemp and Boynton 1984).

Secondary production of coastal ecosystems is also relatively large compared to other biomes of the world (Odum 1971). Biomass production of certain benthic suspension-feeding bivalves in estuarine ecosystems exceeds the highest protein yield of pond cultured herbivorous fish and rivals the areal production of highly subsidised cattle farms (Odum 1967). These remarkable production rates for estuarine animal tissue result, again, from the natural energy subsidy associated with hydrodynamic transport of food and wastes to and from sessile benthic animals. Thus, food chains associated with benthic communities in these relatively shallow estuaries are likely to be more efficient in producing animal tissue. The omnivorous diets of many estuarine animals, and particularly their ability to grow on combinations of living plant material and detrital (dead plant and animal) foods, also result in relatively high trophic efficiencies in organic rich estuaries (Odum 1971). Indeed, compared to freshwater ecosystems, the relative yield of fish per unit of primary production is considerably greater in coastal marine ecosystems (Figure 3.5). This appears to be associated primarily with the mechanical boost associated with physical transport in

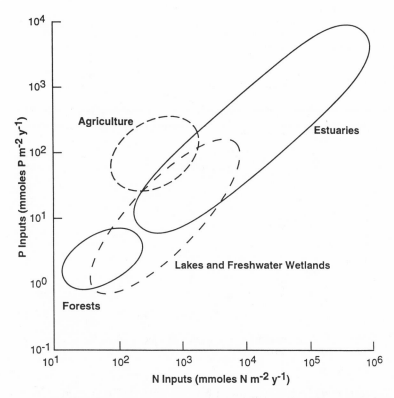

Figure 3.3. Nitrogen versus phosphorous inputs for various ecosystem types. Data are from LMER Coordination Committee (1992).

estuaries. There is, therefore, considerable potential for production of human foods in coastal ecosystems (e.g., Ryther 1969), but the same physical processes (and their associated variabilities) which provide estuaries with natural work subsidies, leading to high trophic efficiencies, also impair the ability of humans to cultivate the coastal seas.

In summary, we find that estuaries and other shallow coastal ecosystems represent unique biomes in which variabilities in certain physical properties (e.g., salinity) and processes (e.g., water movement) are relatively large and unpredictable. As a consequence of natural environmental stresses and the mobility afforded by buoyancy and hydrodynamic transport, taxonomic diversity in estuaries tends to be low with few endemic species. Estuarine functional diversity, however, appears to be relatively high compared to that in other biomes, especially in benthic dominated subsystems. Rates of primary production of coastal ecosystems are among

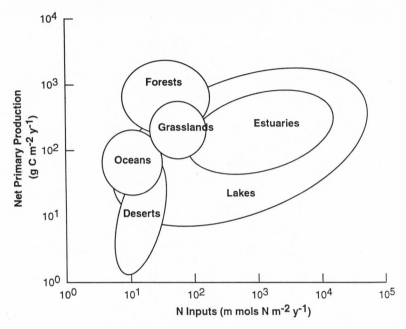

Figure 3.4. Nitrogen inputs versus productivity for various ecosystem types. Data are from Kelly and Levin (1986).

the highest in the biosphere, and trophic transfer of this production to growth of animal populations is relatively efficient. These high conversion efficiencies of sunlight to plant tissue to animal biomass in estuaries are attributable largely to the mechanical subsidy of hydrodynamic processes.

3.3   Resilience and keystone processes in estuarine ecosystems

It has been suggested that many ecosystems exhibit resilient responses to perturbations by developing mechanisms which allow them to "absorb, buffer, or generate change" (Holling 1986). In contrast to stable ecosystems which may oscillate predictably about an equilibrium without much apparent change, a resilient ecosystem is characterised by large fluctuatio..s in response to perturbations. Catalysed, for example, by fire or severe climatic events, a resilient ecosystem may undergo dramatic changes, yet persist. It is these random perturbations that introduce resilience.

In this context, ecosystems contain key organisms and processes which play crucial roles to insure long-term resilience by modifying the impact on ecosystem structure resulting from environmental changes. One mech-

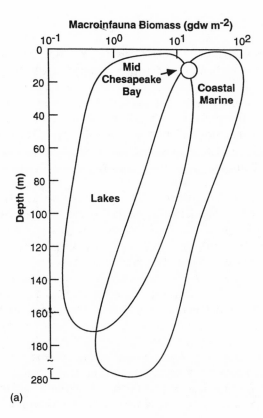

(a)

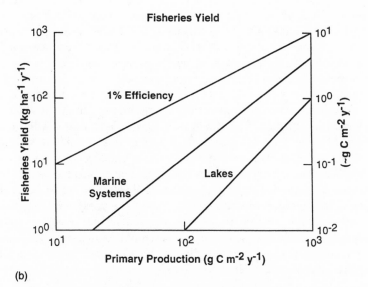

(b)

Figure 3.5. Biomass distribution and primary production versus fisheries yield for marine versus freshwater systems. Data are from Nixon (1988).

anism for conferring ecosystem resilience is to establish alternating replacement structures which are switched periodically but which avoid accumulations of excessive (difficult to replace) structure. Examples of such "keystone" species or processes have been cited for a variety of terrestrial ecosystems. For example, periodic outbreaks of the spruce budworm serve to release accumulated ecological structure for temperate coniferous forests by decimating balsam fir. The outbreaks leave spruce, white birch, and seedlings of fir and spruce, and eventually return the forest to mature stands where fir dominate once again (Holling 1973).

Few if any examples of these kind of key organisms have been identified for estuarine ecosystems. The term "keystone" organisms has also been used to describe organisms that play pivotal roles in the trophic structure of an ecosystem (Paine 1966). In this case, predation by keystone organisms effectively preserves community structure by relieving competitive pressure between organisms at lower trophic levels or serving to keep lower level predation levels in check (Carpenter 1988). This latter kind of keystone organism or process has been identified for shallow benthic ecosystems in coastal marine environments, but they are poorly documented for estuarine plankton systems.

In spite of the sparseness of keystone organisms in estuaries, there is strong evidence that these ecosystems are relatively resilient to perturbations, at least for time scales less than 10^3 years. Estuarine ecosystem processes have been shown to return to predisturbance levels within months after perturbations from major meteorological events – e.g., hurricanes, floods, droughts, and winter freezes. For example, the Chesapeake Bay experienced a 200-year event in June of 1972 when a tropical storm created an unprecedented flood in the lower watershed. Sediment loading to the Bay during the 4–5 days of the storm were equivalent to inputs from the previous decade. Sufficient freshwater was delivered to remove virtually all seawater from the estuary for several days. Despite this dramatic change in environmental conditions, plankton community production and abundance returned to prestorm levels within 2–3 months. Although there were massive mortalities of benthic faunal populations, they recovered within a year. Similarly, there were no apparent effects on the annual fisheries yields comparing 1972 with previous and subsequent years. Numerous other examples are well documented illustrating estuarine ecosystem resilience to major disturbances. A few coastal ecosystems, such as coral reefs, are much less resilient to environmental changes (Jackson 1991), but organisms in these systems are not adapted to the same kind of high-frequency/high-amplitude variations in physical environmental conditions as are those in estuaries.

It appears that this well buffered disturbance-response displayed by various estuarine ecosystems rests on three primary factors:

- the relatively small standing ecological structure;
- the high degree of organism mobility;
- the prevalence of generalist species.

Unlike forests, estuaries and other aquatic ecosystems contain relatively low levels of standing biomass. Hence, the time required to regenerate any losses in ecological structure is relatively short for estuarine systems. The combination of strong hydrodynamic transport and active motility insure that seed organisms are readily available to replace those lost with perturbations. Furthermore, the high degree of functional generalism among estuarine organisms increases the probability that any ecological processes lost with species declines from external perturbations can be replaced by other organisms. In many estuarine ecosystems the rapid plankton turnover resulting from hydrodynamic transport tends to keep any species from dominating the broad ecological niches of estuaries. This tends to facilitate species coexistence and stabilise community structure against internal disruptions such as overgrazing (Kemp and Mitsch 1979).

It appears then that the conventional ecological views of keystone organisms do not apply well for estuarine systems. There are, nevertheless, a number of key organisms and processes which play fundamental roles in the functioning of estuarine ecosystems. Although the loss of these critical ecological attributes may not jeopardise the resilience of an estuarine ecosystem, it may significantly change its ecological structure. Many of these critical processes are most evident in the benthic subsystems of estuaries. For illustrative purposes, we present below an example of a critical process (nitrification-denitrification) which is crucial in nature and is generally not fully recognised until after its loss has occurred. In later sections we give detailed examples of loss in sears populations and decline in oyster, striped bass and shad abundance in the Chesapeake Bay, which reinforce these ideas.

The nitrification-denitrification process

A complex but essential component of nitrogen cycling in estuaries and other biomes of the world is the coupled process of nitrification-denitrification. In this process the reduced nitrogen salt, ammonium, which is released in microbial decomposition and animal excretion is converted to gaseous forms – predominantly di-nitrogen gas (N_2, which composes almost 80% of the earth's atmosphere). Whereas the reduced salts of nitrogen such as ammonium are essential for plant growth, the gaseous forms of nitrogen are virtually unavailable for use by estuarine plants. Under normal conditions, the rate of nitrification-denitrification is directly proportional to the rate of nitrogen loading to the estuary (Seitzinger 1988), so that this coupled process buffers the ecosystem, maintaining an intermediate level of internal nitrogen. In all biomes of the world, including estuaries, nitrifier bacteria are highly specialised organisms, with strict nutrient requirements, especially for oxygen and ammonium, necessary for growth. Thus, nitrifiers are an exception to the rule of few specialist organisms occurring in estuarine ecosystems. Because ammonium concentrations are highest in sediment porewaters, the highest rates of nitrification tend to be concentrated in surface sediments in the narrow zone into which oxygen penetrates. In contrast, denitrifier bacteria are generalists capable of numerous alternative metabolic processes for growth. To metabolise via denitrification pathways, however, they require an abundance of the nitrogen salt produced by nitrifiers (nitrate) as well as total absence of oxygen. Therefore, denitrification will not occur without nitrification and the critical coupling of nitrification and denitrification occurs only at the interface between anaerobic and aerobic environments (Henriksen and Kemp 1988).

In certain estuaries, such as those with deep channels, a high ratio of river flow to tidal flow, and/or limited turbulent mixing, excessive inputs of nitrogen from the watershed can lead to depletion of oxygen from bottom waters. This results from nitrogen-stimulated growth of planktonic algae and decomposition of algal matter near the benthic surface at rates which consume oxygen faster than reoxygenation from the air-water interface. The anoxic conditions which ensue result in massive mortality of bottom-dwelling animals and elimination of any aerobic bacterial processes (e.g., nitrification) which might otherwise occur at the sediment surface. Hence, whereas under normal conditions a significant portion (approximately 50%) of the nitrogen entering the estuary is removed via nitrification-denitrification, under these eutrophic conditions of excessive nitrogen loading and anoxia, this nitrogen removal process is inhibited (Kemp et al. 1990). Within months after oxygen is restored to the Bay bottom, rates of nitrification-denitrification return nearly to the levels occurring prior to the anoxic disturbance. This is, therefore, another example of a key estuarine process which is susceptible to change and which controls the nature of the ecosystem, but which is quickly restored once the perturbation is removed.

3.4   The Chesapeake Bay and its watershed

Because of the special characteristics of the Chesapeake Bay and its watershed, it is, at one and the same time, extremely: (1) productive; (2) unpredictable; (3) resilient; (4) sensitive to stress; and (5) hard to understand and manage with traditional methods. It is productive because it is a broad, shallow, large estuary, where fresh water and nutrients running off a large watershed interact with sea water in very complex and unpredictable patterns. Its shallowness means nutrients are recycled easily and constantly, maintaining its high productivity. Its large watershed means there were always plenty of nutrients even before European settlers converted so much of the watershed to farms, towns, and cities. Because its watershed is so large and it already had plenty of nutrients in its pre-European settlement state, it was very productive and attracted many settlers, but it was also sensitive to the further addition of nutrients that settlement caused. For most of the last 300 years it has been a completely open access resource, leading to its overuse and abuse. Because of its complexity and unpredictability it is still a relatively resilient ecosystem, and if we can come up with ways to intelligently manage its use and reduce the stresses we put on it, it can recover and remain resilient and productive for a long time to come.

We can only do this by first understanding the history and current status of the bay and its watershed as an integrated system, involving the interaction of humans and a unique ecological life support system. We also have to better understand human institutions, how they typically fail when dealing with resources like the Chesapeake and how to fix them or replace them with better alternatives.

A typical estuary

As discussed above, one of the characteristic features of estuaries such as the Chesapeake Bay is that their environments are variable and relatively unpredictable, and the variability occurs on both large and small time and space scales. A simple example is the normal seasonal variation in temperature experienced by these temperate systems. Temperatures range from near 30° C in surface waters in late summer to zero by late winter. Ice cover is extensive in severe winters, but of only local importance in normal winters. Salinity is an important biological parameter and it also varies widely, from near zero in the upper reaches of tributaries to 30 ppt near the capes. Superimposed on this general salinity gradient, significant interannual variability also occurs related to normal climatic shifts from wet to dry periods. As a result, the spatial extent of various estuarine habitats undergoes periods of expansion and contraction.

Shifts from wet to dry periods also strongly influence the rate at which essential nutrients (such as nitrogen, phosphorus and silica) enter the system from the surrounding watersheds. Since these elements are essential for plant growth, their availability determines, at least in part, the amount of organic matter available to support food webs of the bay. Recent measurements indicate that nutrient loading rates and algal production rates vary among years by at least a factor of two.

In addition, the amount of freshwater entering the system is the primary factor determining the degree to which much of the bay system is vertically stratified with lighter, fresher water near the surface and saltier, denser water near the bottom. The biological significance of stratification lies in the fact that the degree of stratification determines the ease with which essential gases such as oxygen can reach deep waters and support the respiratory needs of benthic communities. In turn, the degree of stratification also regulates the ease with which nutrients released from benthic communities reach euphotic surface waters and are again available to support plant growth.

Finally, catastrophic events, such as hurricanes and severe tropical storms, cause yet another, largely unpredictable, scale of variability. These storms dump huge amounts of freshwater, nutrients and sediments into the bay and, through a variety of mechanisms, subject most organisms to some degree of stress. The "memory" of the bay with respect to these events seems to be relatively short (< 5 years) at least for lower levels of the food web. For higher trophic levels we do not know what the long term impact might be.

It is not surprising that estuarine organisms have evolved a variety of physiological and behavioural mechanisms to deal with rapidly varying estuarine conditions. And, it is also not surprising that the number of truly "resident" estuarine species is limited. Thus, despite variable conditions, estuaries still retain identifiable characteristics in terms of species assemblages. Many species exhibit cosmopolitan traits, moving from one geographic or depth zone of the estuary to another (or leave the estuary for part of the year) either as part of their normal life cycle or to avoid unfavourable conditions. For example, during each year many adult striped bass (*Morone saxatilus*) spend a portion of the summer in coastal waters, the winter in deep waters of the bay and the spring in the tidal-freshwater portions of tributary rivers. Small scale vertical movements also occur during summer in response to prey distributions, water temperatures and oxygen conditions.

Other organisms, especially those without the mobility of fish and larger crustaceans, have developed physiological mechanisms to deal, often for extended periods of time, with adverse environmental conditions. The American oyster (*Crassostrea virginica*) has often been cited as a premier example of an estuarine organism. The oyster can grow well across an extreme salinity range, feeds successfully on a broad range of algal species and detritus, stops feeding during cold periods of the year when food supplies are limited and can survive extended periods of hypoxia or anoxia by closing tightly and switching its metabolism to a form of anaerobic respiration. On the scale of the whole estuary, reproduction is favoured in some years in one location and in other locations at other times. Pelagic larval stages insure wide dispersal and colonisation of available habitats and replenishment of areas that have become depopulated. The oyster is but one example of the robust nature of most estuarine organisms.

Finally, estuaries are truly open systems and this has implications for maintenance of species assemblages as well. Both the ocean and landward end of these systems are open to active and passive migrations of both indigenous and exotic species. For example, estuaries are characterised by having both anadromous (ocean dwelling but spawning in estuaries) and catadromous (freshwater dwelling but spawning in seawater) species. In addition exotic species find their way into these systems attached to commercial ships or recreational boats. The constant exchange of fresh and salt waters insures a constant seeding of planktonic organisms as well. Thus, despite the rigorous environment typical of estuaries and intense fishing pressure and degraded habitats in some instances, local extinctions are relatively rare and generally of short duration. Normal migrations, passive entry via river and tidal water flows and accidental introductions insure a continual supply of normal and new species.

3.5   A stressed ecosystem: case studies of declines of critical organisms and processes

Because the Chesapeake Bay has been so productive, some have called it a food factory, or compared it to a powerful engine that runs on nutrients. But the Bay is not a factory or an engine; it is an ecosystem. Instead of machinery, the Bay is composed of living parts: animals, plants and microorganisms that depend on each other. Take away or change some of these living parts, and the whole ecosystem feels the effects – feels them in both direct and indirect ways; ways that are sometimes obvious and sometimes impossible to predict; ways that depend on the subtle interplay of the parts of the system at several spatial and temporal scales that determine whether it is healthy and resilient, or brittle to the breaking point.

Qualitative reports of Chesapeake Bay made during the seventeenth century through the middle of the present century clearly indicate that living resources, in the forms of fish, shellfish and water fowl, were indeed abundant and played an important role in the economy of the region and presumably the ecology of the Chesapeake Bay. For example, William Penn in the late 1600s noted the extreme abundance of seafood as well as the huge size of oysters in the bay; in 1884, annual oyster harvests reached an historical peak of 20 million bushels; the writer H. L. Mencken noted in 1940 that "Baltimore lay very near the immense protein factory of the Chesapeake Bay, and out of it ate divinely."

In the last few decades, reports concerning various fisheries and habitats of the bay have not been generally as positive, and often there have been calls for drastic action to rehabilitate the living resources of the bay. In addition, the habitat diversity of the bay appears to have been greater prior to the last few decades. Some 13 species of submerged aquatic vegetation ringed the bay shores from the tidal fresh rivers to the high salinity waters near the mouth of the bay; the water column was reasonably clear with sunlight penetrating to several meters in most areas and deeper in the more saline portions; oyster reefs provided important topographical relief on the broad shoals of the bay; cooler and deeper waters in the natural channels of the bay provided a refuge from high summer temperatures for a variety of finfish. In recent decades, a considerable fraction of these habitats have been lost. While massive efforts are currently underway to restore bay fisheries and habitats, there have been serious losses of both during the post–World War II period.

While there have been many changes in the status of commercially and recreationally important species in the bay region, the cause-effect relationships involved in these changes are not particularly clear for many. In this section, a brief overview is provided for three commercially important species (striped bass, American shad and American oyster) and one group of species which provide important habitat in the bay (submerged aquatic vegetation). These were selected because they are important in both the economy and ecology of the bay region and because the reasons for declines are reasonably well understood.

Submerged aquatic vegetation communities

It is generally recognised that submerged aquatic vegetation (SAV) communities play an important role in the functioning of shallow water ecosystems in estuaries as well as other aquatic ecosystems. Specifically, studies conducted over the last decade in estuarine systems indicate these systems maintain water clarity in shallow areas by binding sediments and baffling near-shore wave turbulence, modulate nutrient regimes by taking up nutrients in spring and holding these nutrients until fall, and enhance food-web production by supplying organic matter and habitat conducive for rapid growth of juvenile organisms. Submerged plant communities contributed significantly: to food production for Bay fish, invertebrates and waterfowl populations; to habitats used by small animals for refuge from predation; to stabilisation of sediment processes; and to the rapid and efficient cycling of important chemical elements (Kemp and Boynton 1984).

In much of Chesapeake Bay, SAV communities (which include some 13–15 species) started to undergo a serious decline during the 1960s in the upper portions of the Bay (Figure 3.6a) and in the early 1970s in the middle reaches of the Bay (Figure 3.6b). This decline was not taken seriously until the late 1970s when a series of studies were conducted to investigate potential causes. These experiments included field observations, small (50–700 l) and large (400 m³) microcosm exposure tests and simulation modelling and were conducted using several different plant species. Results indicated that the decline was primarily the result of nutrient overenrichment attributable to a combination of pollutants arising *from* human activities in the watershed (Kemp et al. 1983). It appeared that epiphytic algae (a normal part of the SAV community) were overstimulated by enhanced nutrient availability which leads to increased shading of SAV leaves and photosynthetic rates depressed below those needed for healthy plant growth. Increased water column turbidity and adhesion of suspended sediments to SAV leaves further reduced available light. Herbicides were found to be a relatively small factor in the decline although in areas of the bay adjacent to agricultural drainage, seasonal herbicide stresses were possible (Kemp et al. 1983). As a consequence, there was a dramatic shift in the Bay's ecosystem trophic structure from a system of balanced production via benthic and planktonic food chains to one dominated by plankton. Although the estuarine community structure has changed markedly with the loss of these plant communities, overall productivity of the Bay has not been diminished. Another measure of the estuary's resilience to this major disturbance lies in the fact that many plant species have returned rapidly (within 5 years) to their former abundance levels in the one Bay tributary in which pollutant loading has been substantially reduced. The efficient hydrodynamic transport of plant propagules from clean environments landward and seaward of the estuary facilitated rapid recovery once the cause of the disturbance had been removed. So it appears that if nutrient loading rates to these systems are reduced, SAV communities are capable of re-establishing themselves in many areas of the bay (USEPA 1982).

American oyster

The oyster is a gregarious animal; it prefers to grow in groups. Because of this attraction, oyster larvae set and grow in clusters, ultimately forming large aggregations called oyster bars (or rocks or reefs).

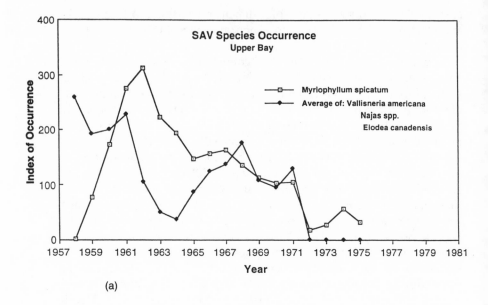

(a)

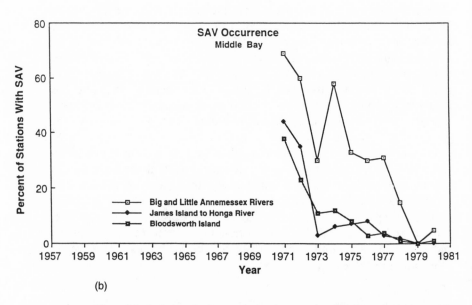

(b)

Figure 3.6. An index of submerged aquatic vegetation abundance in the lower (a) and middle (b) salinity regions of the Chesapeake Bay for several time periods. Data are from Bayley et al. (1978) for the upper Bay and from the U.S. Environmental Protection Agency (1983) for the middle Bay.

These days, after years of harvesting by oyster tongs or dredge, an oyster bar may lie low and scattered across the bottom of the Bay. But during the Colonial period, when Captain John Smith first sailed the Bay, the oyster bars were said to reach from the Bay bottom all the way to the water's surface. These bars actually formed oyster reefs, and like coral reefs in the tropics, oyster reefs undoubtedly created diverse ecosystems.

Imagine, for a moment, oyster reefs stretching up both sides of the Chesapeake Bay, along the shallow margins. Fish and other marine animals would have gathered around the reefs to feed, and chances are that one could see these fish, because the water would be relatively clear. The water during those early years would be less murky than now for two reasons. First, because prior to European settlement and the introduction of intensive agriculture, the land surrounding the Bay and its rivers was covered with forests, forests that protected the soil and prevented runoff. And second, the Bay would be more transparent because the oysters themselves were actually cleaning the water.

Oysters are filter feeders. An oyster pumps about 50 gallons of water a day in order to filter out algae (also called phytoplankton), the tiny floating plants that serve as its primary food source. As oysters feed, they act like filters in a swimming pool, drawing out algae and clearing the water. Roger Newell has estimated that the Bay once had so many oyster reefs that the oysters could pump through a volume of water equal to the entire Chesapeake Bay in less than a week (Newell 1988). Because oyster populations have dwindled to such low levels, it would now take a year or more for today's oysters to filter that same amount of water. Their decline probably contributed to the loss of SAV described above.

The long-term record of commercial oyster harvests for Maryland and Virginia portions of the bay are shown in Figure 3.7 for the period 1929–89. From 1929 through about 1960 combined state catches fluctuated between 20 and 40 million pounds per year. From 1960 through the early 1980s combined catches decreased to 20–25 million pounds per year with virtually all of the decrease occurring in Virginia waters. However, there was a rapid decline in waters of both states beginning in 1981 and this trend has persisted through the present time. There is considerable debate within the scientific community as to the relative importance of several factors in causing this decline and there is equal if not more intense debate in management agencies concerning possible actions to rebuild this resource. Whatever management actions may eventually be taken it appears that overfishing, disease and loss of habitat have been the principle factors responsible for the decline of this resource.

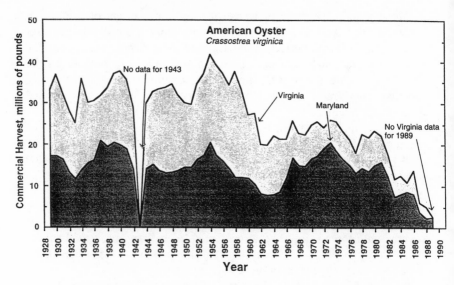

Figure 3.7. Annual commercial catches of American oysters from Maryland and Virginia waters between 1929 and 1990. Data are from Jones et al. (1990).

Striped bass and American shad

The annual commercial catches of striped bass and American shad from both Virginia and Maryland portions of Chesapeake Bay for the period 1929–90 are shown in Figure 3.8a and 3.8c. In addition, an index of striped bass spawning success in Maryland waters is shown in Figure 3.8b. Striped bass catches have undergone a multi-decade period of increase followed by a sharp decline in the last decade. At the present time there is a relatively strict ban on striped bass fishing with only brief and highly regulated seasons in the spring and fall. Fishing for American shad is closed in the Bay and has been closed since the early 1980s. As with other commercially important species in the Bay, there is debate as to the most important causes of these declines. In the case of striped bass, reduction in spawning stock size (due to overfishing) and habitat degradation appear to be the most likely causes. The fishing ban on striped bass has yielded increased stock sizes but the regular pattern of successful recruitments every 2–4 years has yet to reappear. The high juvenile index observed in 1989 was largely restricted to one location in one tributary river and it is doubtful that it represents a Bay-wide recruitment success. Despite the fact that the fishing ban on American shad has been in effect longer than the ban on striped bass, there is little indication that the stock has started to rebound. One reason for this lack of response is that shad normally migrate much farther upstream than striped bass before spawning. Since virtually all tributary rivers of the bay are dammed, it is generally assumed that blocked access to optimal spawning areas has been a major factor impeding the reestablishment of this stock.

3.6   Summary of impacts on living resources

In recent years an increasing amount of public attention, research and management planning have been directed towards rehabilitating living resources in Chesapeake Bay as well as other estuarine and coastal marine systems. In this regard the Chesapeake is certainly not unique. Declines, some of them very severe, have been noted above for several important recreational and commercial species.

While the causes of these declines have been much debated, and the degree of certainty regarding cause-effect linkages varies widely, there is general agreement that the following factors have generally been involved, either singly or in combination, in declines in living resources in the bay. Records of commercial *fishing pressure* on most stocks in the bay are poor in some cases and nonexistent in most. In addition, there are very few records of recreational fishing effort but it is believed to be in excess of commercial effort for some species. Management actions designed to reduce fishing pressure, and thus rehabilitate a stock, have been successful in some instances and less so in other cases. In recent years the availability of reliable *water quality* information has increased sharply. These data are currently being examined to determine if there are correlative relationships between success of particular species and water quality status of regions of the bay. In particular, poor dissolved oxygen conditions in deep waters are thought to be a likely factor impacting some benthic stocks and certain fish species. In addition, an examination of the role of *toxic* compounds has also been initiated. It appears clear that there are toxic "hot spots" in the bay system and these are adjacent to several industrial centres (e.g., Norfolk, Virginia and Baltimore, Maryland). Within these regions there are clear indications of impact (e.g., deformed fish, skin lesions) at the level of individual fish. However, the impact at the population and community levels of organisation remains largely unknown. It does appear that toxic stresses operate at the sublethal level in areas other than the grossly impacted areas referred to previously.

In response to changes in water quality, there have been some potentially important changes in the *habitat diversity* of the bay as well. Over the past 20 years submerged aquatic plant communities have been greatly reduced in the bay, now occupying only 5–10% of previous sites. These communities have been shown to be important feeding and refuge areas

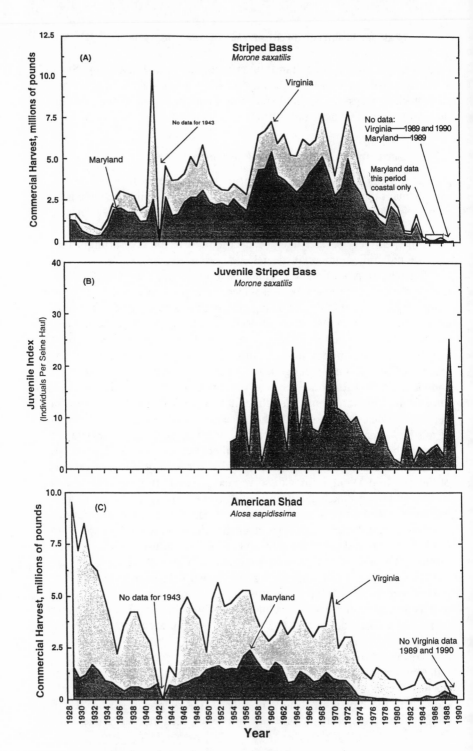

Figure 3.8. Annual commercial catches of striped bass (a) and American shad (c) from Maryland and Virginia waters between 1929 and 1990. Also shown is an index of striped bass spawning success (b) for the Maryland portion of the Bay. Data are from Funderburk et al. (1991).

for a number of species and have also been shown to play a significant role in modulating nutrient and sediment cycling in littoral regions of the Bay. In deep waters, oxygen conditions have also deteriorated thus removing a cool water refuge for finfish during the warm summer months and removing the entire benthic habitat for several months each year. Finally, *diseases* have played a fairly well documented role in the decline of some species of both direct and indirect commercial importance. In recent years the remaining oyster stocks have been severely impacted by two diseases. In the 1930s one of the important sears species (*Zostera marina*) was attacked by a fungal disease which severely reduced the distribution of this species in the higher salinity portions of the Bay.

Human population growth and change in the Bay watershed

During the Revolutionary War, when George Washington travelled through Annapolis, about 500,000 people lived in Maryland. In the two centuries that have passed since, Maryland's population has grown almost ten times. Now 4.7 million people live in the state, and according to predictions by the Maryland Office of Planning, another 800,000 or more will settle in Maryland by the year 2020. Maryland's growth illustrates change throughout the Chesapeake watershed. Today, more than 14 million people live in the watershed; that number is expected to top 17 million by the year 2020.

Population changes in the watershed are shown in Figures 3.9a–d. The number of dots is proportional to population in each county and distributed randomly within each county. Between 1940 and 1986 the population of the watershed increased 87%. About 20% of this increase was due to net migration into the watershed, the majority of it to the areas surrounding Baltimore and Washington in Maryland and Virginia. The population of the watershed grew at an average annual rate of 1.6% between 1952 and 1972, almost the same as the U.S. average of 1.5% for the same period. But the growth was concentrated in the Maryland and Virginia portions of the watershed, which averaged 2.6% growth compared to the remainder of the watershed which grew more slowly at only .4%.

The most striking changes were in three areas: Richmond, the Norfolk-Virginia Beach area, and the Baltimore-Washington corridor. Growth can, in part, be attributed to increases in industry related directly and indirectly to the expanding U. S. government as well as the increasing fashionableness of the Chesapeake Bay as a recreation area. The latter forces are amplified by the high immigration rates to the area.

These areas also illustrate the "urban flight–suburban sprawl" phenomenon that has at once undermined the more natural and rural atmosphere

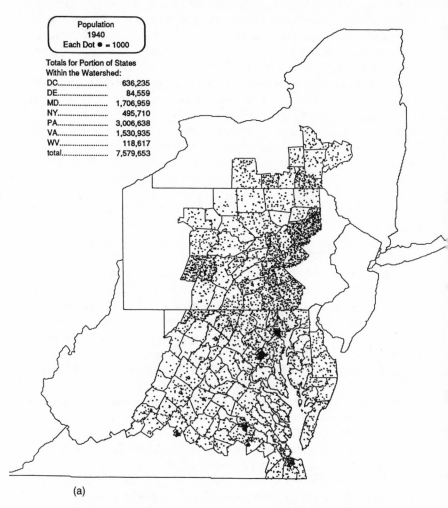

Population
1940
Each Dot ● = 1000

Totals for Portion of States
Within the Watershed:
DC........................ 636,235
DE........................ 84,559
MD........................ 1,706,959
NY........................ 495,710
PA........................ 3,006,638
VA........................ 1,530,935
WV........................ 118,617
total........................ 7,579,653

(a)

Figure 3.9. Population in the Chesapeake Bay Watershed. (a) 1940; (b) 1952; (c) 1972; (d) 1986.

that many originally left the city for, while at the same time removing businesses and middle and upper income residents which served as a revenue base for the cities. The resulting deterioration of services and infrastructure worsens as people move farther and farther out. Increasing travel times required to reach work with the concurrent degeneration of traffic conditions, and soaring property values are some of the resistive forces that quell the further spread of suburban development.

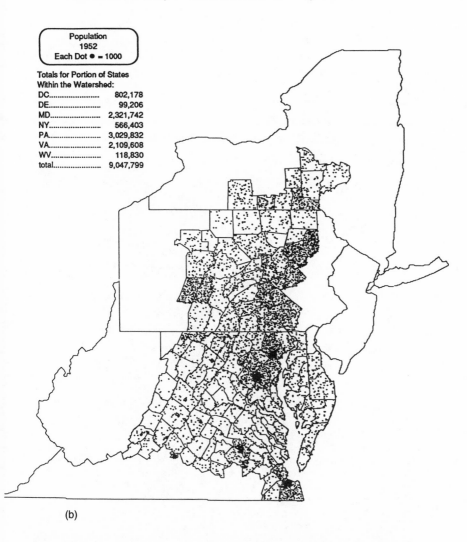

Population
1952
Each Dot ● = 1000

Totals for Portion of States
Within the Watershed:
DC........................ 802,178
DE........................ 99,206
MD........................ 2,321,742
NY........................ 566,403
PA........................ 3,029,832
VA........................ 2,109,608
WV........................ 118,830
total........................ 9,047,799

(b)

It appears that these forces may in fact be approaching an equilibrium, at least for the time being. Between 1972 and 1986 the growth rate of the Maryland watershed population slowed to an annual rate of .9%, and in Virginia to 1.8%, and emigration rates from the cities have slowed. For example Washington D.C.'s emigration rate decreased from an annual average of 1.3% in the 1960s to less than .8% in the early 1980s. In addition the spatial extent of the sprawl appears to be presently limited to the counties immediately surrounding the cities in question (Figure 3.10). However, the information contained in these maps is only suggestive. Demographic

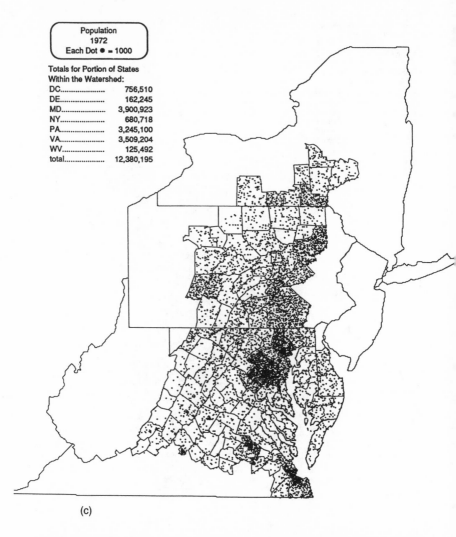

Population
1972
Each Dot ● = 1000

Totals for Portion of States
Within the Watershed:

DC...................	756,510
DE...................	162,245
MD...................	3,900,923
NY...................	680,718
PA...................	3,245,100
VA...................	3,509,204
WV...................	125,492
total...................	12,380,195

(c)

trends can be the result of any number of factors. Variation in birth rates, cultural heritage, local versus long distance moves, political climate and zoning laws all muddy the waters. In addition, as sprawl and growth occur simultaneously, secondary economic centres inevitably spring up initiating their own cycles.

As population has risen in the watershed, both land and water have felt the effects. Increases in evidence of human activity accompany the increases in population in both magnitude and distribution. Maps of manufacturers, energy consumption, housing units, water use, solid waste pro-

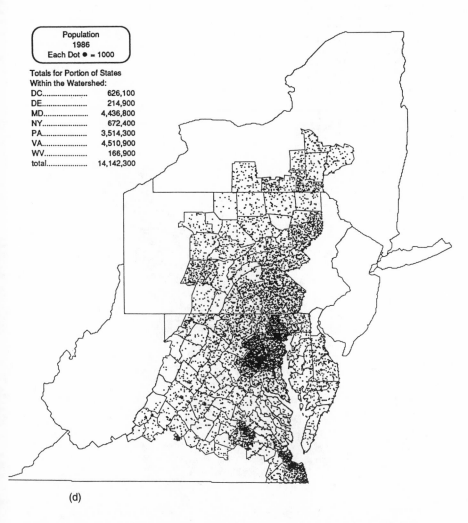

Population
1986
Each Dot ● = 1000

Totals for Portion of States
Within the Watershed:
DC...................... 626,100
DE...................... 214,900
MD...................... 4,436,800
NY...................... 672,400
PA...................... 3,514,300
VA...................... 4,510,900
WV...................... 166,900
total................... 14,142,300

(d)

duction, and air pollutants all closely resemble the maps of population. However, changes in lifestyle have caused accelerated increases in consumption and waste production. From 1952 to 1986 we have seen the following increases in the watershed: *per capita* energy consumption has gone from 567,000 BTU/day to 744,000 BTU/day; NO_x emissions by vehicles has gone from 1.0 lb/week per person to 1.7 lb/week per person; and per capita solid waste production has risen from 2.2 lb/day to 3.7 lb/day. In contrast public and industrial per capita water use has decreased in the same period for the watershed as a whole – from 334 gal/day to 278 gal/

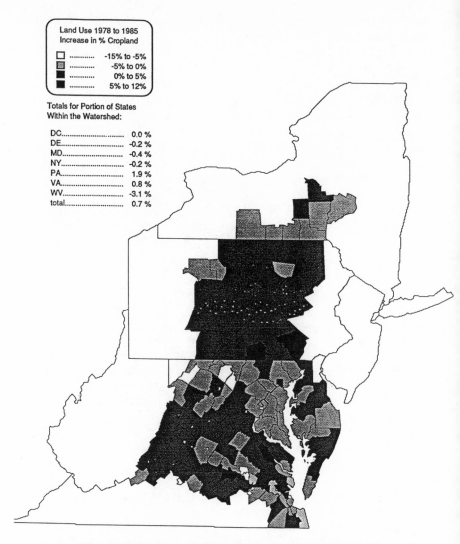

Figure 3.10. Percentage change in urban land use from 1978 to 1985 in the Chesapeake Bay watershed.

day. This is due in large part to the decline in total use in Pennsylvania during this period from 1.46 billion gal/day in 1952 to only 1.18 billion gal/day in 1986. In the Maryland and Virginia portions of the watershed the per capita use rate held nearly constant at about 237 gal/day. This suggests that the former change may be due to changes in heavy industry in Pennsylvania, while the latter lack of change may be due to a relatively constant per capita public demand for water in a region which is primarily residential and light industry.

Agriculture

Changes associated with agriculture are tied to increases in population in their overall magnitude, and to cultural and historical practices in their distribution and local magnitude. Heavily populated areas necessarily exclude agriculture. It is unfortunate that some of the best agricultural land in the country, which was the basis for the initial local growth, is rapidly being converted to residential developments, industrial parks, and shopping malls as rising property values make farming unprofitable. It is not clear, however, whether agriculture itself is more benign to an ecosystem such as the Chesapeake Bay than urban and suburban development. Both load the system with wastes and nutrients while consuming "natural" areas that could have absorbed some of that load.

While total farm acreage has decreased from 23.2 million acres in 1954 to 14.4 million acres in 1987 in the watershed, cropland has only decreased from 10.5 million acres to 9.0 million. Meanwhile from 1954 to 1987 the average farm size increased from 126 acres to 190 acres. This means larger percentages of farms are devoted to crops (64% vs. 45%) while less lies fallow, pastured, and wooded. Operations are larger and more intensive. Irrigation has increased from 40 thousand acres to 180 thousand acres, and fertilisation rates have increased from about 210 lbs per cultivated acre to 250 lbs per cultivated acre. Figure 3.11a–c shows pesticide use which was almost nonexistent in the early 1950s, peaked in the late 1970s, and has decreased from 1974 when 15 thousand tons were used, compared to 13 thousand tons in 1986. Of course this also reflects the greater specificity of the pesticides used in 1986.

If we examine the application of fertiliser nitrogen (Figures 3.12a–c) we see a pattern which follows the general distribution of farmland but with an overall increase in the intensity of use. While the phosphorus content of fertilisers used in the watershed remained relatively constant over the years at around 10%, the nitrogen content steadily rose from about 7% in the 1950s to 15% in the mid 1980s.

The nutrient loading rates for the Chesapeake Bay system are currently at 151.68 million kg per year nitrogen and 11.25 million kg per year phosphorus (Figure 3.13). As Figure 3.13 shows, more than half of both the nitrogen and phosphorus inputs are from diffuse sources like agricultural and urban runoff. Figure 3.14 also shows the importance of denitrification as a sink of nitrogen, consuming about 26% of the total input.

If one assumes that the loading rate is proportional to the amount of

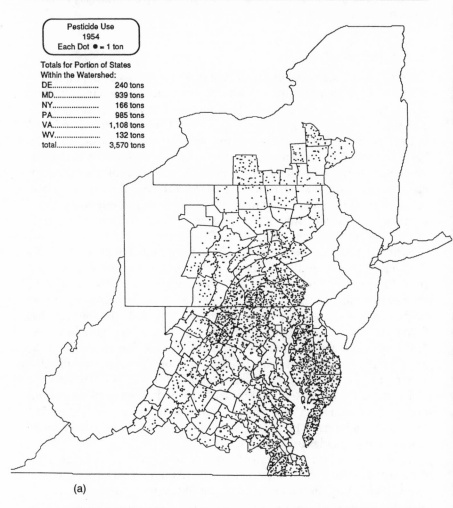

Pesticide Use
1954
Each Dot ● = 1 ton

Totals for Portion of States
Within the Watershed:
DE...................... 240 tons
MD..................... 939 tons
NY...................... 166 tons
PA...................... 985 tons
VA..................... 1,108 tons
WV..................... 132 tons
total................... 3,570 tons

(a)

Figure 3.11. Pesticide use in the Chesapeake Bay Watershed. (a) 1954; (b) 1974; (c) 1987.

nutrients produced and applied within the watershed, then this implies that phosphorus loading has increased around 4 to 10% and nitrogen loading has increased somewhere between 35 to 125% since 1954. This would not only be a great change in amount of nutrients but (and this may be even more crucial) a great change in the relative proportion of nutrients. The increased concentrations of nutrients in specific localities

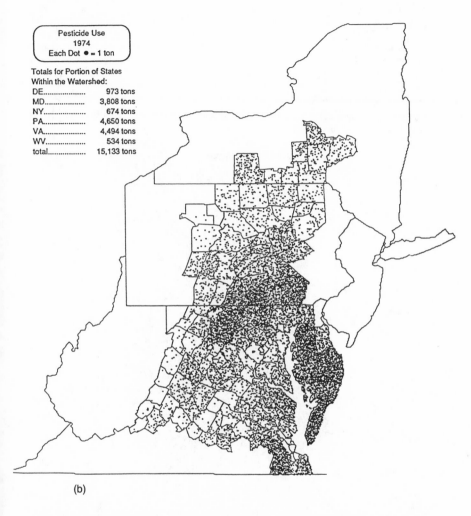

Pesticide Use
1974
Each Dot ● = 1 ton

Totals for Portion of States
Within the Watershed:
DE....................	973 tons
MD..................	3,808 tons
NY.....................	674 tons
PA.....................	4,650 tons
VA.....................	4,494 tons
WV....................	534 tons
total...................	15,133 tons

(b)

such as Lancaster County means that this problem may be even more exaggerated in some parts of the system.

Summary

The Chesapeake Bay has undergone very rapid population growth with its associated environmental impacts. We have mapped some of these changes as they are reflected in the characteristics of the Bay's watershed (see Table 3.1). The impacts of these activities in the watershed and on the

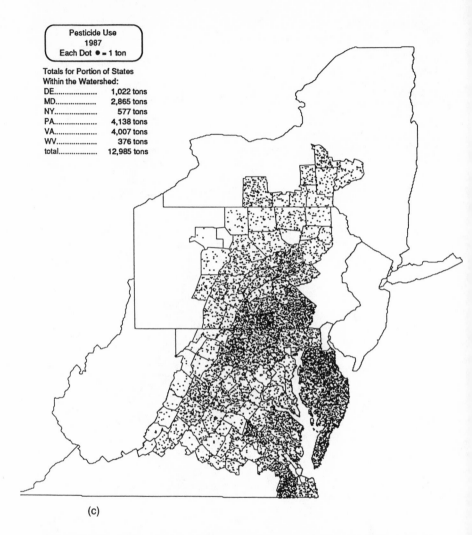

Pesticide Use
1987
Each Dot ● = 1 ton

Totals for Portion of States
Within the Watershed:

DE....................	1,022 tons
MD....................	2,865 tons
NY....................	577 tons
PA....................	4,138 tons
VA....................	4,007 tons
WV....................	376 tons
total..................	12,985 tons

(c)

Bay itself are known to be large, but their specific interconnections are only now being investigated. The Chesapeake Bay has 200,000 people living in its drainage basin for every km³ of water in the Bay (the Baltic Sea has 4,000 people/km³, and the Mediterranean has 85 people/km³ by way of comparison). Even if all of these people were minimising their environmental impacts (which they are not) their sheer numbers are daunting to a system as sensitive as the Chesapeake. If these numbers continue to increase as they have been in the past the prospects for America's largest estuary seem bleak.

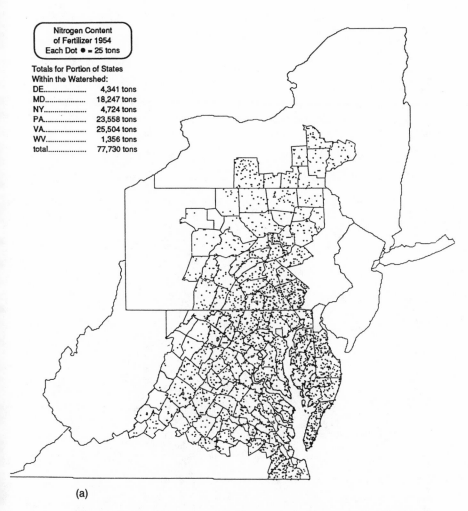

(a)

Figure 3.12. Nitrogen content of fertilizer use in the Chesapeake Bay Watershed. (a) 1954; (b) 1974; (c) 1987.

3.7   Biodiversity and economic values of Chesapeake Bay and other estuaries

Given our previous discussion, it is clear that estuarine systems like the Chesapeake Bay have high economic value for a number of reasons, but their biodiversity is generally not one of them. Estuaries in general have rather low biodiversity, but are nonetheless very productive and resilient

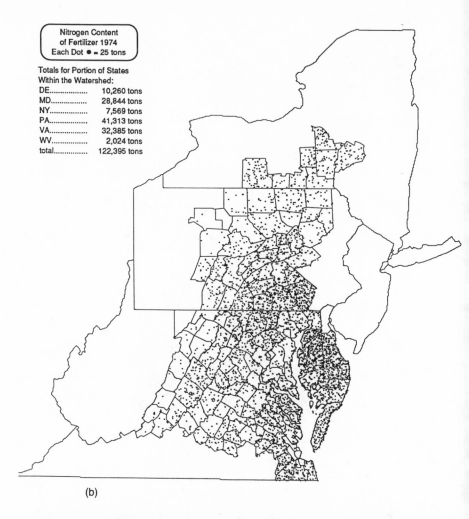

Nitrogen Content
of Fertilizer 1974
Each Dot ● = 25 tons

Totals for Portion of States
Within the Watershed:
DE.................   10,260 tons
MD.................   28,844 tons
NY.................   7,569 tons
PA.................   41,313 tons
VA.................   32,385 tons
WV.................   2,024 tons
total.................   122,395 tons

(b)

ecosystems. Because of the dominance of unpredictable physical forces in estuaries and the lack of a stable base on which to build biological structure to smooth out this unpredictability, these systems are dominated by a relatively few generalist species. The economic value of estuaries focuses on these few species. For example, in the Chesapeake the American oyster, striped bass, American shad, and blue crab have been primary economically important species for both commercial and recreational fisheries (Cumberland 1988, 1989, 1990). Their dramatic decline in recent years has been attributed to a combination of overfishing and mismanagement of

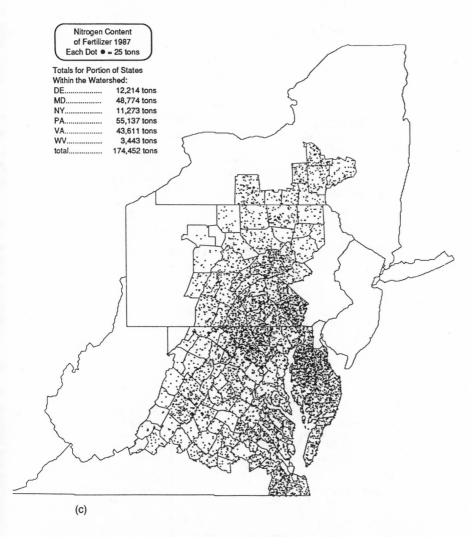

Nitrogen Content
of Fertilizer 1987
Each Dot ● = 25 tons

Totals for Portion of States
Within the Watershed:
DE.................. 12,214 tons
MD.................. 48,774 tons
NY.................. 11,273 tons
PA.................. 55,137 tons
VA.................. 43,611 tons
WV.................. 3,443 tons
total.............. 174,452 tons

(c)

two key processes in the Bay, namely, submerged aquatic vegetation and nitrification-denitrification as mediated by oxygen and nutrient input levels.

Toward a theory of scale and biodiversity

Given the foregoing discussions and insights about biodiversity in estuaries, we have formulated an embryonic hypothesis about the relationship of biodiversity to the scale and predictability of the environment. This

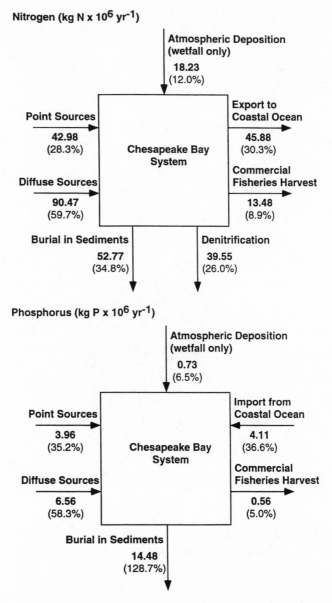

Nitrogen (kg N x 10^6 yr^{-1})

Atmospheric Deposition
(wetfall only)
18.23
(12.0%)

Point Sources
42.98
(28.3%)

**Chesapeake Bay
System**

Export to
Coastal Ocean
45.88
(30.3%)

Diffuse Sources
90.47
(59.7%)

Commercial
Fisheries Harvest
13.48
(8.9%)

Burial in Sediments
52.77
(34.8%)

Denitrification
39.55
(26.0%)

Phosphorus (kg P x 10^6 yr^{-1})

Atmospheric Deposition
(wetfall only)
0.73
(6.5%)

Point Sources
3.96
(35.2%)

**Chesapeake Bay
System**

Import from
Coastal Ocean
4.11
(36.6%)

Diffuse Sources
6.56
(58.3%)

Commercial
Fisheries Harvest
0.56
(5.0%)

Burial in Sediments
14.48
(128.7%)

Figure 3.13. Nitrogen and phosphorus budgets for the Chesapeake Bay system as of 1992.

Table 3.1. *The watershed's population at a glance*

State or district	1940	1986	2020 (projected)
District of Columbia	636,235	626,100	626,100
Delaware	84,559	214,900	215,000
Maryland	1,706,959	4,436,800	5,496,600
New York	495,710	672,400	700,000
Pennsylvania	3,006,638	3,514,300	3,854,500
Virginia	1,530,935	4,510,900	6,229,800
West Virginia	118,617	166,900	166,900
Total watershed	7,579,653	14,142,300	17,288,900

Note: Only those portions of each state within the watershed are included.
Source: Bureau of the Census, 1952, 1988, Maryland Office of Planning.

hypothesis may be useful in sorting out the value of biodiversity, both to the ecosystems themselves and to human consumers of ecosystem services.

In the discussion that follows, we employ the "4-box model" of Holling (1987, 1992a). Holling proposes four basic functions common to many ecosystems and a spiralling evolutionary path through them, expanded here to encompass many complex systems including economic systems (Figure 3.14). The functions (boxes) are: (1) *Exploitation* (r-strategists, pioneers, opportunists, entrepreneurs, etc.); (2) *Conservation* (K-strategists, climax, consolidation, rigid bureaucracies, etc.); (3) *Release* (fire, storms, pests, political upheaval, etc.); and (4) *Reorganisation* (accessible nutrients, abundant natural resources, etc.).

Within this model, systems evolve from the rapid colonisation and exploitation phase, during which they capture easily accessible resources, to the conservation stage of building and storing increasingly complex structures. Examples of the exploitation phase in ecology are early successional ecosystems colonising disturbed sites or, in economics, pioneer societies colonising new territories. Examples of the conservation phase are climax ecosystems or mature, large bureaucracies. The release or "creative destruction" phase represents the breakdown and release of these mature structures via aperiodic events like fire, storms, pests, or political upheavals. The released structure is then available for reorganisation and uptake in the exploitation phase.

The amount of ongoing release or creative destruction that takes place in the system is critical to its behaviour. The conservation phase can often build elaborate and tightly bound structures by severely limiting creative destruction (the former Soviet Union is a good example), but these structures become "brittle" and susceptible to massive and widespread destruc-

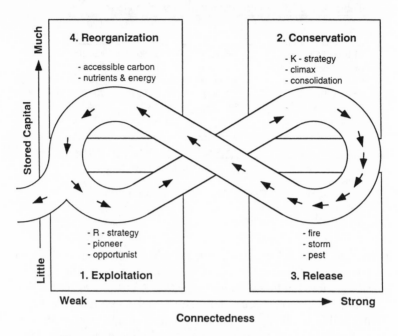

Figure 3.14. The Holling "4-box" model.

tion, as the former Soviet Union again illustrates. If some moderate level of release is allowed to occur on a more routine basis, the destruction is on a smaller scale and leads to a more resilient system. Creative destruction, in terms of shocks or surprises, seems to be crucial for system resilience and integrity. Similarly, it has been argued that episodic events, such as the Chernobyl accident, the Rhine chemical spill, the death of seals in the North Sea, are shocks to the social-cultural value system and may stimulate change towards more resilient ecological economic systems (Folke and Berkes 1992).

In terrestrial ecosystems, fire climax systems such as the pine forests of Yellowstone National Park are good examples of creative destruction. In its unmanaged state, Yellowstone burned over extensive areas relatively often, but because of the high frequency the amount of fuel was insufficient to allow highly destructive fires. The more frequent, small to moderate size fires would release nutrients stored in the litter and support a spurt of new growth, without destroying all the old growth. On the other hand, if fires are suppressed and controlled, fuel builds up to high levels and when the fire does come (because control and suppression are *never* perfect), it wipes out the entire forest. As an analogy in economics, the demise of the former Soviet Union after decades of control and stability is equally illustrative.

Estuaries, in this context, are always awash in creative destruction due to the strong physical forces of tides and currents that dominate this unpredictable environment. They are constantly reset to the exploitation phase and never build up enough structure to make it to the conservation phase. They are resilient, low taxonomic diversity systems. Only systems with relatively predictable environments can build and maintain a diverse set of specialist species. Tropical rain forests are the extreme case of both environmental predictability and biodiversity. In this view a stable environment *allows* biodiversity to develop, rather than the reverse argument, that high biodiversity leads to a more stable ecosystem response.

But the process is seen as a feedback loop. Ecosystems in unpredictable environments at one scale can build structures at a larger scale to smooth out and stabilise that unpredictability. In estuaries the scale of the unpredictability is the same as the scale if the organisms and estuaries cannot build large biostructure and therefore remain at low diversity. An exception that proves the rule is the case of artificial structures like bridge pilings, on which diverse biological communities do grow because of the smoothing effect of the artificial structure. In some ecosystems (like forests) structure *can* be built to smooth out lower scale unpredictability. But this process can be tampered with by artificially reducing the amount of release or creative destruction (e.g., the Yellowstone forest fire management policy) and the system can become brittle.

Estuaries do have high functional diversity, however, and high resilience. Figure 3.15 summarises our view of estuaries and their relationship with their physical environment. The top half of the diagram shows the effects of the high variance, low predictability physical forces driving estuarine ecosystem dynamics with high efficiency secondary production and high fisheries yields. The bottom half of the diagram shows the effects of the physical forces and linkages with ecosystem dynamics on ecosystem structure, with low taxonomic diversity, moderate biomass, and high functional diversity. Large and unpredictable physical forces cause structural losses and keep high taxonomic diversity from developing, but also enhance productivity.

From this analysis of estuaries we have developed an hypothesis about the general relationship of resource predictability to scale and biodiversity. The biodiversity (or taxonomic diversity) in a system is a function of the predictability of the resource environment on the time and space scales above, at, and below the scale of the system of interest. All else being equal, the higher the predictability of the environmental resources and forcings, the higher the biodiversity that can develop in an attempt to

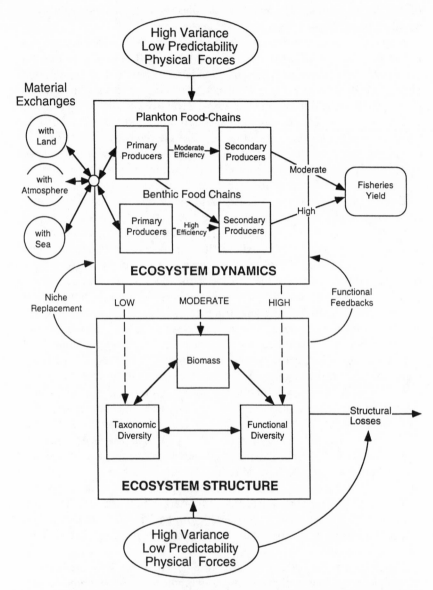

Figure 3.15. Diagram of the relationships between physical forcings and ecosystem structure and dynamics in estuaries.

maximise the efficiency of use of these resources. The absolute amount of biodiversity is limited by the absolute size of the resources and forcings. The system can also be tampered with by artificially restricting the relationship between structure and predictability as in Yellowstone and the former Soviet Union. However, these attempts lead to brittle systems that ultimately collapse.

We think that the above hypothesis is testable using comparative analysis of various system's resource predictability, diversity, and structural dynamics over several scales. It is applicable to both economic and ecological systems and may help us to better understand the role of diversity in systems' performance and value – a key issue in modern conservation policy.

PART II

INTEGRATING ECOLOGY AND ECONOMICS
IN THE ANALYSIS OF BIODIVERSITY LOSS

CHAPTER 4

Wetland valuation: three case studies

R. K. Turner, Carl Folke, I. M. Gren and I. J. Bateman

4.1 Introduction

Should society be concerned about the loss of wetlands? The arguments and analysis in this chapter seek to construct a case for an affirmative answer to this question. Wetlands represent important forms of natural capital (Costanza and Daly 1992), and there is an urgent need for a balance to be struck between wetland conservation, sustainable utilisation and wetland conversion (Turner 1991). The management process will not be costless and therefore wetland assets require proper ecological and economic valuation.

Wetlands are multifunctional and can be considered as capital assets which require appropriate (sustainable) management if they are to continue to produce the flow of wetland derived functions, services and goods. This flow is generated by species and processes interconnected within their environment in what is referred to as life-support systems (Odum 1989). Life-support systems generate a range of ecosystem produced functions, services and goods of fundamental value to society.

Some life-support functions of ecosystems can be valued in economic terms, but others may not be amenable to meaningful monetary valuation. It has been doubted whether the full contribution of component species and processes to the aggregate life-support service provided by ecosystems can be fully captured in economic values (e.g., Ehrlich and Ehrlich 1992). We would further argue that the value of the aggregate life-support service including the ecosystem structure and function has seldom been recognised and not explicitly taken into account in economic calculations. This prior – or primary – value of the ecosystem refers to the buildup, the succession and the evolution of the system structure, which is the very basis for its life-support capacity. Wetland flora and fauna perform a major role in this process.

Wetlands have been and still are degraded in many parts of the world.

129

This chapter starts with a section on the current status of the world's wetlands and causes of wetland loss. The third section discusses the role that wetland species play in ecological processes and functions. The fourth and fifth sections deal with wetland valuation. We claim that it is the secondary values of wetland ecosystems, and only a minor part of them, which have been dealt with in the economic analyses and valuations. In section five we provide three wetland valuation studies and put them in the context of primary and secondary values. The chapter ends with a discussion on the significance and value of wetlands in relation to the valuation studies and to a sustainable use of natural capital. From the sustainability perspective, it is important to make sure that wetland values are properly reflected in, and integrated into economic policy.

4.2   Current status of the world's wetlands and causes of wetland loss

Wetland ecosystems account for about 6% of the global land area and are among the most threatened of all environmental resources. Since 1900, over half of the world's wetlands may have disappeared. The temperate wetlands in developed economies have long suffered significant losses and they continue to face an on-going conversion threat from industrial, agricultural and residential developments, as well as from hydrological perturbation, pollution and pollution-related effects.

Although areas such as North America and Australia retain significant and relatively pristine wetland stocks, these stocks are nothing like as large as they once were. The United States alone has lost an estimated 54% (87 million ha) of its original wetlands, of which 87% has been lost to agricultural development, 8% to urban development and 5% to other conversions (Maltby 1986). In most of Europe, the amount of wetlands acreage that remains is only a fraction of the original stock, and is often close to, or below, critical threshold levels. Precise loss estimates on a national basis are not generally available. Several OECD member countries did implement policy changes, beginning in the mid-1970s, designed to slow down wetland conversion but the impact of these policy changes has not been sufficiently monitored for a clear judgement to be made about their success. Nevertheless, there is a strong probability that wetland losses continue to be high in many OECD countries (Turner and Jones, 1991).

The total area and status of tropical wetlands are still unknown, but the available evidence suggests that the pattern of wetland conversion in developing economies mimics the North American and European temperate wetland loss process – it may even be proceeding at a faster rate in some regions of the developing world (Barbier 1989a, 1989b, 1992). Mangrove swamps, for example, are rapidly disappearing through Asia and Africa, because of land reclamation, fishpond construction, mining and waste disposal (Dixon 1989). Countries such as Bangladesh, Cameroon, Chad, India, Niger, Thailand and Vietnam have lost over 80% of their freshwater wetlands (UNEP, 1992).

It seems clear, then, that wetlands (of all types) have been disappearing at an alarming rate around the globe. All wetlands are under threat from a variety of locally or regionally based human activities. But currently the perilous state of coastal wetlands is under an additional threat posed by global-scale pollution emissions and consequent global warming. The forecasted accelerated increase in average sea levels would have serious impacts on the worldwide distribution of coastal wetlands. Salt, brackish and freshwater marshes in temperate zones, and mangroves and other swamps in tropical zones would be inundated or eroded (Turner, Kelly and Kay, 1990).

Three types of interrelated "failure" phenomena can be identified as playing an important role in this process:

1. *Information failures* – a general lack of appreciation of the full economic value of conserved wetlands, which has contributed to subsequent market and intervention failures.

2. *Market failures* – in particular, wetlands have suffered from pollution (externality problem); and from water supply diminution as a result of excessive abstraction of open-access water resources (public good failure). Another form of the externality problem occurs when wetland benefits do not accrue to the wetland owner, resulting in a divergence between private and social benefits.

3. *Intervention failures* – the general absence of properly integrated resource-management policies has resulted in intersectoral policy inconsistencies and wetland degradation. There are also examples of inefficient policies directed at wetlands themselves, and of policies directed at other sectoral issues which carried with them unintended "spillover" effects for wetlands.

All of the historical loss of wetlands does not necessarily represent "inefficient" resource use. This is particularly so in developing countries, where the multiple use of wetlands can lead to significant improvements in social welfare. However, loss of wetlands, particularly large scale losses, implies loss of species diversity. Species that go extinct are irreversibly lost. The social value of avoiding the loss of species is discussed in Perrings et al. (1992), and the part that species play in generating wetland values in the following section.

4.3   Species and wetland life support functions

Wetlands develop through a combination of changes brought about by species responding to environmental influences and species interacting within the biological community. Hydrological pathways, which transport energy and nutrients to and from wetlands, are fundamental for the establishment, and maintenance of these ecosystems and their processes. The hydrological regime influences pH, oxygen availability, and nutrient flux, and these parameters to a large extent determine the development of the biological part of the wetland system. The biological part in turn modifies the hydrologic conditions by trapping sediments, interrupting water flows, and building peat deposits (Gosselink and Turner 1978). Because of the accumulation of peat (soil organic material mainly from plants which have died) the variability of, for example, flooding is reduced, and a storage of nutrients is built up, providing a steady source of nutrients to wetland plants. Plant species composition and primary productivity of wetlands are controlled by the duration of flooding, the turnover rate of water, and the quality of inflowing water (Ewel 1991).

The modification and stabilisation of environmental variability, performed by the biological subsystem during the development of the wetland, insulates the ecosystem from its environment, which makes the system less dependent on and affected by external inputs. At the level of single species this occurs through structural and physiological genetic adaptations to mineral salt and sediments that are anoxic, and at the ecosystem level through the production of peat, making nutrient recycling possible (Mitsch and Gosselink 1986). However, the wetland system is an integrated part of the landscape, and its biodiversity and generation of environmental goods and services is therefore affected by what is taking place in adjacent ecosystems, especially if adjacent ecosystems are radically perturbed.

Wetlands are known for their flora and fauna, particularly rare plants and migratory bird species. The composition of species in wetlands varies both in time and between wetland sites. Differences and abundance of species in wetland ecosystems are discussed in e.g., Mitsch and Gosselink (1986), Williams (1990) and Ewel (1991). The presence and abundance of species living in a wetland depends on their life histories, which are often short but complex, and their specific adaptation to the environmental conditions of the wetland site. The wetland environment is in many ways a physiologically harsh habitat. Wetlands in general have low oxygen levels in the water column, anoxic soils, periods of drying and flooding, and high water temperatures at the wetland surface. There is significant sediment-water exchange due to the shallow water in wetlands, and a diversity of decomposing organisms in wetland sediments lead to processes such as denitrification and chemical precipitation which remove chemicals from the water. In general physical and microbial processes are more important than vegetative uptake in controlling sediment and nutrient retention (Johnston 1991).

In the following section we will focus on the life-support functions that wetland ecosystems generate. A major reason for conserving the biological diversity of wetlands is for the role that their species play in generating and sustaining such functions.

4.4 Wetlands as sources of environmental values

Wetlands provide a wide array of environmental goods and services of high value to society. They change sharp runoff peaks, from heavy rains and storms, to slower discharges over longer periods of time, and thus prevent floods. They recharge groundwater aquifers, and provide drinking water in dry seasons. Many wetlands act as sinks for inorganic nutrients and many are sources of organic materials to downstream or adjacent ecosystems. They have a capacity to improve water quality and often serve as a filter for wastes, reducing the transport of nutrients and organic materials, sediments and toxic substances into coastal areas. For this reason they are often referred to as the kidneys of the landscape. Wetlands are also involved in global biogeochemical cycles and contribute to the global stability of available nitrogen, atmospheric sulphur, carbon dioxide, and methane. Furthermore, wetlands are important habitats for flora and fauna, and serve as nursery and feeding areas for both aquatic and terrestrial migratory species. For these reasons wetlands often have high recreational, preservational and aesthetic values.

We conceive of "valuing" a wetland as essentially valuing the characteristics of a system. Since it is the case that the component parts of a system are contingent on the existence and functioning of the whole, then putting an aggregate value on wetlands and other ecosystems is rather more complicated a matter than has previously been supposed in the economic literature (Turner 1992). The total wetland, including its dynamic changes over time, can be considered as a source of primary value or "glue" value. The primary value includes the redundancy or "insurance" capacity of ecosystems, for example the emergence of new keystone species and processes as systems respond to unexpected shocks and change over time.

The existence of the wetland structure *is prior* to the range of functions and services which the wetland system then provides. The aggregated value of this flow of wetland system output we term total secondary value. Total secondary value is conditional on the continued "health" of the ecosystem. The use of one ecological service, one secondary value, often implies that other secondary values are affected. Each secondary value is dependent on the existence, operation and maintenance of the multifunctional wetland system – the life-support system (Folke 1991 a,b).

In this ecological economic framework, the often used concept of total economic value, which includes use and nonuse value, has at least two limitations. First, it may fail to fully encapsulate the total secondary value provided by an ecosystem, because in practice some of its functions and processes are difficult to analyse (scientifically) as well as to value in monetary terms. Secondly, the concept of total economic value fails to capture the primary value of ecosystems, indeed the "glue" or life-support value notion is very difficult to measure in direct value terms since, it is a non-preference, but still instrumental, type of value.

We believe that this suggestion of a primary and secondary ecosystem value classification goes some way towards satisfying many scientists' concerns about the "partial" nature of the conventional economic valuation approach. It is also a classification that avoids the instrumental versus intrinsic value in nature debate which we believe has become rather sterile. A more formal classification of primary and secondary values in relation to the total economic value concept is given in Turner, Doktor and Adger (1994).

4.5   Empirical case studies of wetland values

In Sections 4.6, 4.7 and 4.8 we synthesise three case studies of wetland evaluation. They reflect different ways of quantifying the primary and secondary values of existing, degraded and restored wetlands. The first is a comparison of the lost support functions value of a Swedish wetland and the costs, monetary and biophysical, of trying to replace this support with human-made technologies. It attempts to indirectly estimate both primary and secondary wetland values. The second is an application of the contingent valuation method to the Broadland wetlands, United Kingdom. It attempts to elicit recreation and amenity values from a large sample of wetland visitors and nonvisitors. The study therefore estimates a part of the total secondary value of the wetland. The third is a policy analysis study relating to the island of Gotland, Sweden, and represents an attempt to combine biophysical and economic measures. The study estimates the value of wetland restoration for the purpose of nitrogen abatement, taking into account the multifunctionality of the wetland support, the temporal dynamics of restored wetlands functioning as nutrient sinks, the efficiency of nitrogen cleansing in relation to other abatement technologies, and the effects on the aggregate income of the region. The study therefore addresses the notion of primary value and estimates secondary values.

4.6   Replacements costs and biophysical evaluation of life-support functions

The first case study, based on Folke (1991b), is an attempt to quantify the life-support value of a Swedish wetland system, the Martebo mire, on the island of Gotland in the Baltic Sea. The mire has been subject to extensive draining and most of the wetland-derived goods and services have been lost.

The purpose was to evaluate the life-support functions of the wetland to society. This was done by comparing the loss of the wetland's functions with the costs of replacing them, where feasible, with human-made technologies. The ecological functions and services of the wetland and the human-made replacement technologies are summarised in Table 4.1.

It is difficult to analyse the significance of life-support systems solely in monetary terms because their functions seldom have a market and the general public seldom has information about the support functions of ecosystems. For these reasons the evaluation of life-support functions in this study is mainly founded on a biophysical perspective.

In ecology, the amount of energy captured via photosynthesis (gross primary production, GPP) is a common measure of an ecosystem's potential to generate environmental goods and services. The loss of the life-support, or the primary value of the wetland, was approximated via an analysis of how much of the capacity of wetland plants to capture the sun's energy had been lost. The annual loss corresponded to about 730 TJ of GPP (unexploited wetland 1725 TJ GPP per year – exploited wetland 995 TJ GPP per year). The evaluation of the costs for human-made technologies necessary to replace lost component secondary values generated by the life-support system prior to its destruction was made through monetary estimates, and industrial energy estimates.[1] Using energy analysis is *not* to argue for an energy theory of value. The energy approach as applied here is a complement to economic analysis, since it might help to reveal interrelations not reflected in monetary valuations, and also assist in the identification of a biophysical framework within which the market may operate. The strengths and weaknesses of the methods are thoroughly discussed in Folke (1991b).

The results of the study are summarised in Table 4.2. They indicate that it takes considerable amounts of costly industrial energy to replace the

---

[1] Fossil fuels, electricity etc. used directly and indirectly in society to produce goods and services.

Table 4.1. *Life-support functions, environmental goods and services of the Martebo mire, exploitation effects, and replacement technologies*

Societal support	Exploitation effects	Replacement technologies
Peat accumulation	Peat layer reduction /disappearance through decomposition, intensive farming, wind erosion, degraded soil quality	Artificial fertilisers redraining of ditches
Maintaining drinking water quantity	Reduced water storage Lost source for urban area	Water transport[1] Pipe line to distant source[2]
Maintaining ground-water level	Dried wells	Well drilling Saltwater filtering[2]
Maintaining drinking water quality	Saltwater intrusion Nitrate in drinking water Pesticides in drinking water	Water quality controls Water purification plant Silos for manure from domestic animals nitrogen filtering[2] Water transport, dams for irrigation, pumping water to
Maintaining surface water level	Decreased evaporation and precipitation, reduced amount of water	dam Irrigation pipes and machines Water transport for domestic animals[1]
Moderation of waterflows	Pulsed run-offs Decreased average water flow in associated stream	Regulating mire Pumping water to stream
Processing sewage, cleansing chemicals	Reduced capacity Eutrophication of ditches and stream	Mechanical sewage nutrient and treatment, sewage transport, Sewage treatment plant, clear-cutting of ditches and stream nitrogen reduction in sewage treatment plants[2]
Filter to coastal waters	Adding to eutrophication	
Providing		Agriculture production
• Food for humans	Loss of food sources	Imports of food[3]
• Food for domesticated animals	Loss of food sources	Roof materials[3]
• Roof cover	Loss of construction materials	Releases of hatchery raised trout[4]
Sustaining		
• Anadromous trout populations	Degraded habitat commercial and sport fishery losses	Farmed salmon[4]
• Other fish species	Loss of habitat	
• Wetland dependent flora/fauna	Loss of habitat	
Species diversity	Endangered species	
Storehouse for genetic materials	Lost	
Bird watching, sport fishing, boating and other recreational values	Lost	
Aesthetic and spiritual values	Lost	

[1]These replacements take place in other drained areas on the island of Gotland
[2]These replacements are being planned
[3]Not considered further in the analysis because of difficulties in quantifying the losses of food sources and construction materials
[4]There is still natural reproduction in the stream, but there are also occasional releases of hatchery raised trout, the cost of farming salmon is a measure of the commercial losses of the reduced number of returning adult trout.
Source: Modified from Folke (1991a).

Table 4.2. *Undiscounted monetary costs and industrial energy costs of technologies aiming at replacing the lost wetland support in thousands of 1989 Swedish kroner, and in fossil fuel equivalents TJ ($10^{12}J$)*

Replacement technologies	Monetary costs	Energy costs
Redraining and clear-cutting of ditches and stream (g)	50 - 56	0.200 - 0.330
Dams for irrigation	57 - 205	0.200 - 1.167
Pumping water to dams	11 - 36	0.025 - 0.225
Irrigation pipes and machines (h)	58 - 184	0.230 - 1.045
Artificial fertilisers (g)	935 - 1345	8.130 - 19.095
Regulating wire	10 - 18	0.060 - 0.200
Pumping water to stream (h)	7 - 11	0.015 - 0.070
Well drilling	33 - 53	0.080 - 0.300
Water quality controls	12 - 40	0.020 - 0.180
Water purification plant	32	0.180 - 0.230
Nitrogen and saltwater filtering (h)	0 - 460	0 - 3.290
Water transports (h)		
• Humans	0 - 500	0 - 4.050
• Domestic animals	0 - 335	0 - 2.715
Pipe line to distant source (h)	0 - 990	0 - 5.570
Mechanical sewage treatment and storing (g)	625 - 750	1.360 - 4.265
Silos for manure from domestic animals (g)	370 - 950	1.160 - 5.405
Sewage transport	63 - 126	0.165 - 1.020
Sewage treatment plant	3 - 200	0.005 - 1.170
Nitrogen reduction in sewage treatment plant (g)	40 - 45	0.085 - 0.255
Hatchery raised trout		
Farmed salmon (b)	210 - 250	0.720 - 1.670
Endangered species (b)	68 - 314	0.120 - 1.965
Total	2585 - 6900	17,950 - 46,865

[1]Technologies attributed to loss in the performance of biogeochemical processes (g), the hydrological cycle (h), and to the biological part (b) of the wetland ecosystem; these performances are interdependent, however.
Source: Modified from Folke (1991a).

loss of wetland produced goods and services. These goods and services were previously generated "for free" by the wetland system. Most of the costs, whether monetary or biophysical, are related either to technical substitutes for the biogeochemical processes in the wetland, or to substitutes for the services associated with the hydrological cycle. Less than 10% are directly related to the biological part of the system. This might reflect a lower priority and also the difficulty of substituting for the loss of species diversity and genetic variability (irreversibility).

A monetary estimate of the replacement cost (discounted and undiscounted) does not provide us with a precise measure of the total biophysical life-support value of the wetland. This is because the estimate of the monetary replacement costs tells us only that the annual (undiscounted) cost of replacing the wetland's functions is about 2.5–7 million SEK ($.4–1.2 million). We have estimated some of the secondary values and part of the system's support value but not the total value. In the context of a wetland conservation vs. development situation we might compare this estimate with the benefits of alternative uses of the wetland, but we would still have an underestimate of the actual total support value of the wetland. This is especially critical if the alternative uses being appraised are based on unsustainable options such as intensive agriculture (Daly and Cobb 1989).

There is a need for complementary estimates of the physical foundations and interrelations between ecosystems and economic systems. The energy analysis applied here provides a biophysical indicator of the lost life-support functions to society. It is an attempt to also highlight the loss of primary value, or glue value as discussed above. The loss of individual components of the life-support service was compared with the costs of technical replacements. This was done by comparing the industrial energy used throughout the economy to produce and maintain the replacement technologies with the solar energy required by the wetland system to produce and maintain similar ecological functions. The analysis indicated that the biophysical cost of producing technical replacements in the economy (15–50 TJ fossil fuel equivalents per year) was almost as high as the loss of life-support functions, measured as solar energy fixing ability by plants (55–75 TJ fossil fuel equivalents per year). However, this is not an "equivalent" substitution because the human-made replacements for the wetland's life-support functions have actually not succeeded in restoring the original ambient environmental quality, there are still severe environmental problems in the area.

Hence, according to this study's results technical replacements only represent a partial substitute for elements of the ecosystem support, and only then at a high cost, since expensive and environmentally degrading fossil fuels are used directly and indirectly in the economy to produce and maintain the replacements. From a societal perspective, it would be much wiser to use these "replacement fuels" in a prior effort to maintain and enhance the free support of ecosystems, instead of trying to replace it with human-made technologies after it has been destroyed. One solution along these lines would be to apply ecological engineering (Odum 1971; Mitsch and Jörgensen 1989) to restore wetlands for drinking water supply, recreation, and as filters for nitrogen to coastal waters (Ewel and Odum 1984; Nichols 1983). The economics of wetland restoration will be analysed in the third case study below.

4.7   Broadland: a contingent valuation study

Broadland represents a "class one" complex wetland because it is recognised as providing a wide range of functional and structural values. Broadland is, therefore, of considerable significance for wetland conservation. There are three National Nature Reserves within it – Bure Marshes, Hickling Broad, and Ludham Marshes, the first two of which appear on the list of wetlands recognised by the British government (under the Ramsar Convention), as being of international importance. In addition, there are 24 SSSIs, with more due to be so designated in the near future. The Broadland fens support reed and sedge beds and companion fauna. The grazing marshes support important wildfowl populations, and their drainage dykes contain over a hundred different species of freshwater plants. Broadland was specifically designated an "Environmentally Sensitive Area" under the Agriculture Act of 1986.

The Broadland wetlands also provide for multiple resource use and encompass multiple land-use systems. The area is a national centre for recreation. It supports a regionally important agriculture industry; a substantial permanent population; as well as a large seasonal tourist population. The origins of the Broads lie in the flooding of medieval peat diggings, in the connection of these shallow lakes to the main water courses, and in the creation of a marsh-based economy. Today, this seminatural area supports a variety of intensive and extensive agricultural land uses on the pump-drained marshland. The combination of chalky soils and agricultural activities yields a significant impact on the chemical composition of the water draining into the waterways. The result is that the lower reaches of the river and some downstream Broads are both affected by saline intrusions. Relatively small changes in rates of sewage effluent discharge, or in water extractions (for example, for spray irrigation and residential consumption purposes) can quickly alter water quality levels. Undrained wetlands are located on peat soils in the upper part of the valleys, while the drained marshes and arable lands lie below the level of the waterway, usually behind clay-peat embankments.

The management of Broadlands has proven to be a complex task, and several factors have helped to intensify land-use conflicts over recent years. Increasing demands on the productive use of wetland resources – such as for increased agricultural output; for more recreation; and for more water for consumption and/or effluent assimilation – have created undesirable side-effects. There has been an accelerated enrichment of the watercourses by nutrients (eutrophication), which has, in turn, led to algal growth and decay, and the associated loss of vegetation, and to organic decay. Changes in the characteristic landscape of the area have also been stimulated; loss of reed banks; channelisation and quay-heading of river banks; loss of grazing marsh and of related dyke habitats; and loss of natural ornithological and invertebrate attributes (Turner and Jones 1991).

Broadland remains under continued threat from flooding, some twenty thousand hectares lying below the surge tide level. This area is protected by over two hundred kilometres of tidal embankments many of which are old and in a deteriorating condition. By the mid-1980s, various pieces of scientific evidence had become available, all of which indicated that the risk of flooding in the Broadland area was probably increasing over time. In the aftermath of major flood damage in 1953, a substantial investment was made in flood protection works for the entire east coast of England. Most of these works have a maximum design life of 30–50 years, and many existing defences are now reaching the end of their useful life. The current standard of flood protection provided by the river walls in Broadland is frequently below a one in ten (or even a one in five) year flood return period.

On the assumption that some overall level of flood protection (from saline inundation) for Broadland is essential, cost-effectiveness analysis, rather than cost-benefit analysis, is now being used to plan a *selective flood protection strategy*. The strategy has multiple goals, and encompasses an economic analysis which is being run in parallel with environmental impact investigations, in order to determine priority areas for protection. Selectivity is therefore interpreted in a number of dimensions – economic, environmental and aesthetic/ethical. Within the overall strategy, the costs of alternative means of achieving given levels of flood protection are being compared with damage-cost-avoided measures of benefits for agricultural, residential/industrial and environmental assets.

Most recently, a contingent valuation study (CVM) has been undertaken to try to assess the monetary value (willingness to pay; WTP) of conserving the Broads via a protection strategy designed to mitigate the increasing risk of flooding due to the long term deterioration of flood defences (Bateman et al. 1992). More specifically the CVM aimed to examine the value of conserving the largely nonmarket assets of recreation and environmental quality as currently provided by the wetland complex. A "do nothing" scenario was formulated to enable survey respondents to judge for themselves the relative merits of the current wetland asset structure and the changed environment subject to frequent flooding events.

The CVM study contained both an on-site survey of users and a mail questionnaire sent to nonusers. Both groups were asked for their WTP to conserve Broadland in its present condition. The on-site survey utilised two different types of questionnaire. One subsample was presented with an open ended (OE) format WTP question – "what are you willing to pay?" – and mean WTP was calculated as a simple average across the sub-sample. A second subsample was initially presented with a dichotomous choice (DC) format WTP question, "are you willing to pay £x?" with the value x being systematically varied across the subsample. The DC mean WTP was calculated via a discrete variable modelling process. The respondents in this latter sample were then allowed to revise their bids up or down from the amount x in an iterative bidding (IB) process. This allowed the calculation of a third IB mean WTP estimate.

The on-site research strategy was to use these three estimates of mean WTP to produce an "envelope of valuation" encompassing "true" WTP. Theory and earlier studies suggest (Bateman and Turner 1992) that the opportunities for "free-riding" (individuals understating the full amount they would be WTP for a good) presented by the OE approach would produce a downwardly biased result and the DC approach an upwardly biased result. These two results were then taken as the respectively lower and upper bounds for the WTP estimate range. Furthermore the IB approach, by starting with the (upwardly biased) DC sum X and then allowing respondents to fix their own WTP (i.e., to free ride downwards), was expected to produce a mean WTP in between those derived from the OE and DC experiments.

The on-site survey instrument contained two forms of information provision – a "constant information statement" which was read out by the trained interviewers to each respondent; and a visual display information board which was on hand at each interview point. The information board was a largely pictorial display of the current Broadland landscape and other asset features (the "before" pictures and likely conditions with frequent flooding the "after" pictures).

The survey questionnaire was extensively tested prior to the main survey using a large pilot sample (> 400 interviews). In total, the whole survey process was undertaken at 17 survey sites located throughout Broadland and involved more than 3,200 completed interviews. The main findings of the pilot survey were that the general taxation payment vehicle outperformed all other vehicles (charitable donation, trust, etc.) and that the range of DC bid levels for the main survey should be set according to the results of the OE WTP questions in the pilot survey (probable underestimate). The DC bid levels were therefore set at £1, £5, £10, £20, £50, £100, £200, and £500.

Main on-site survey results indicated insignificant hypothetical bias problems. Refusals accounted for only 1% of the OE sample and 4.5% of the DC/IB sample. Part-whole (or mental account) bias was assessed by asked respondents to state their annual recreation/environment budget. This was then compared to their stated WTP to conserve Broadland.

Table 4.3. *Annual aggregate user value for Broads conservation (millions of pounds at 1991 prices)*

	Forecast annual visitation	
Elicitation technique	Broads authority (upper bound)	Countryside commission (lower bound)
DC (upper bound)	25	19
OE (lower bound)	8	6

Note: All results rounded to nearest £1 million.
Source: See text.

Analysis indicated that there was no evidence of any significant part-whole bias, with stated WTP coming out at around 16% of the annual budget for both sub-samples. However, it is likely that the well perceived nature of Broadland amongst visitors and the fact that it is a unique U.K. resource (i.e., it is both the part and the whole) contributed significantly to the result that was obtained. A checking procedure in the questionnaire also ensured that stated WTP did really represent annual rather than once-and-for-all payments.

Mean OE (WTP) came out at £77 per household p.a. and IB at £84 (1991 prices). Given the unique status of the Broadland complex these results are roughly in line with other UK CVM studies. Thus a recent study of river water quality (many substitute goods) recorded a mean OE WTP of just over £12 per annum (1987 prices). A study investigating the conservation value of The Yorkshire Dales (mountain and valley landscape area with only a few equivalent substitute sites) found a mean OE WTP of £35 per annum (1989 prices). The signs and significance of explanatory variables such as income, age, first or repeat visit, membership of environmental pressure group etc. were all satisfactory and in accordance with economic theory.

The mean DC (WTP) result was £244 per household p.a. on a bid function that proved robust and consistent with theory. Further testing confirmed the existence of a strong upward anchoring bias in DC WTP responses. Table 4.3 presents a range of user value estimates for the Broads wetland. The results are differentiated both in terms of the elicitation method used and the forecasted annual visiting rate for the area that was used.

In addition to the on-site survey of users, a mail survey across Britain was also used. The objective of this latter survey was to estimate conservation (existence-type values) values held by nonusers. Both socioeconomic factors and a distance (from the Broads area) decay factor were tested for in the stated WTP results.

Only the distance-decay relationship appeared to be significant with those respondents living in a defined "Near-Broadland" zone having a higher WTP (£12.45 per household producing an aggregate annual nonuse value of £32.5m for the zone) than those living in the "Elsewhere GB" zone (£4.08 per household producing an aggregate annual nonuse value of £77.3m for the zone). However these results did not adequately distinguish between past users of the Broads and pure nonusers and cannot therefore be classified as pure nonuse values.

The contingent valuation study just described has examined the value of conserving the secondary wetland values of recreation and amenity, of the Broads wetland complex.

4.8 Gotland: wetland restoration for nitrogen purification

In many countries, the emission of nitrogen has been an important cause of eutrophicated coastal waters and high concentrations of nitrate in ground water. In Sweden, as in several other countries, the search for measures to mitigate these problems has mainly been focused on sewage treatment plants and changes in agricultural practices. However, it is now recognised that other quite different types of measures, implying the use of nature's own nitrogen purification capacity (so called eco-technologies), can be considered as alternatives to the conventional nitrogen abatement measures. Because wetlands function as nitrogen sinks, the restoration of wetlands can be regarded as an important ecological technology for nitrogen abatement. The purpose of this study is therefore to calculate and compare the value of restored wetlands with the value of using conventional technologies involving sewage treatment plants and changed agricultural practices. The study is applied to Gotland, an island in the Baltic, where the level of nitrate in ground water is high due to the load of nitrogen that has been received over time.

The capacity of restored wetlands to purify watersheds (reducing nitrogen and other pollutant levels) has been pointed in several studies carried out by natural scientists; see e.g., Mitsch and Jorgensen (1989) and Nichols (1983). There is, however, still an ongoing scientific debate over the efficiency of wetlands as nutrient sinks (Race 1986). Only a small number of economic studies of the restoration of wetlands have been undertaken. Most of the economic studies have focused on the social costs associated with wetland degradation; see e.g., Bergström et al. (1990), Costanza et al. (1989). To our knowledge, there is only one study where the cost of nitrogen abatement via wetlands restoration has been compared with the costs for using conventional abatement technologies (Andreàsson-Gren, 1991).

There are some similarities in the methods used to find measurements of the value of restoring wetlands and of existing wetlands. A common feature is the difficulty of measuring, not only the value of nitrogen abatement, but also the full multifunctional value provided by the wetland (e.g., water buffering capacity and other several life-supporting functions). There are, however, two further questions that must be explicitly addressed when assessing the value of restoring wetlands. When a certain function of the wetland is of interest, as in this case nitrogen abatement, the value (benefit) of restored wetlands is partly determined by the efficiency of its nitrogen purification compared with other abatement technologies. The other difference concerns the change over time in the capacity of restored wetlands to function as nutrients sinks. When measuring the value of existing wetlands it is usually assumed that the production of different environmental services does not change. However, the capacity of restored wetlands to provide these different services increases over time, as the plants grow and spread, until they reach their mature state. At this stage the wetland has to be harvested to continue to function as a nutrient sink. Before the empirical results from Gotland are presented we give a very brief description of the derivation of the values of nitrogen abatement by the different measures. Unless otherwise stated, all analytical and empirical results referred to are found in Gren (1992).

Valuation of nitrogen abatement measures

Three alternative nitrogen abatement measures included in this study are: restoration of wetlands, expansion of the capacity of sewage treatment plants and a reduction in farmers' use of nitrogen. The benefits of the various nitrogen abatement measures are calculated by applying an intertemporal model where environmental resources are derived from the values of provided outputs as suggested by Mäler (1991a and 1992). The analytical results will now be briefly described.

The value of restored wetlands as nitrogen sinks is derived from all the outputs generated by the life-support system. Thus, in addition to improved water quality other benefits are obtained from wetlands restoration such as water buffering and biodiversity conservation, see Section 4.6. The marginal value of restoring the wetland in time t_1, w_t can then be expressed as follows:

$$W_t = \int_t^\infty e^{-(r-k')(r-t)} (V + A) \, \delta t$$

where r is the discount rate and k' is the natural growth of wetlands. V is the marginal value of water quality and A is the marginal value of ancillary benefits. According to this expression the present value of a marginal increase in the stock of wetlands in time t includes current and all future streams of values from the corresponding increase in water quality and ancillary environmental services. Due to the initial increasing capacity of wetlands to provide these services, the associated values available in the future grow at the rate k'. Note that the future values are discounted by the discount rate, r, minus the growth of wetlands, k'.

The corresponding value of an increase in the stock of sewage treatment plants in time t, S_t, is

$$S_t = \int_t^\infty e^{-(r+\rho)(r-t)} (V)\delta t$$

where ρ is the rate of depreciation. By comparing the W_t and S_t we note that the marginal value of restored wetlands is likely to exceed the value of a marginal increase in the nitrogen abatement capacity at the sewage treatment plants for two reasons. First, future values of a marginal increase in the stock of sewage treatment plants are discounted at a higher rate than the values of an increase in the stock of wetlands, i.e. $(r + \rho) > (r - k')$. Secondly, augmenting the stock of wetlands yields the additional value of ancillary environmental services provided by wetlands.

The value of reducing farmers' use of nitrogen by one unit is simply the corresponding increase in the value of water quality. So, value of reducing farmers' use of nitrogen = marginal value of water quality. So,

$$N = V$$

Thus, according to the analytical results based on the above three expressions, the value of nitrogen abatement via restoring wetlands is likely to exceed the values of nitrogen abatement via the other methods. This is because of the multifunctionality of wetlands and the natural growth in wetlands. In the next section, these values are estimated for nitrogen abatement in Gotland.

Application to Gotland

The most serious environmental problem in Gotland is an insufficient supply of drinking water of acceptable quality, particularly during the dry summer months. The average level of nitrate in Gotland is high, 40 mg NO_3/l, compared to the rest of Sweden at 10 mg NO_3/l. In some wells the level of nitrate exceeds 100 mg NO_3/l. The main sources of nitrogen are the instantaneous leakage of nitrogen from drained mires and farmers application of nitrogen fertiliser and manure. Although the average application rate of nitrogen is modest, about 100 N kg/ha, the leakage of nitrogen is high. This is due to the bedrock structure. The bedrock in Gotland is mainly limestone which is porous thus making it easy for nitrogen to infiltrate and reach the ground water. The bedrock also contains numerous cracks, so that nitrogen is quickly spread from one area to another. For this reason and also due to lack of data, the spatial allocation of nitrogen emission sources is neglected. The ground water basin in Gotland is thus treated as a single recipient.

As shown in the following section, the value of improved water quality was included in the marginal values for all the nitrogen abatement measures. A monetary measure of water quality was obtained from a Swedish study in which the contingent valuation method was used to estimate the willingness to pay for a certain reduction in the level of nitrate in drinking water (Silvander 1991). According to the study's results, the willingness to pay for a reduction from 50 mg NO_3/l to 30 mg NO_3/l was SEK 600/person/year (1 USD = SEK 5.80). This value was therefore used as the upper limit of the valuation function which was then assigned a quadratic form in terms of the level of nitrate.

In order to estimate the production function for water quality a hydrological model of Gotland was used (Spiller 1978). According to the simulation results of this model, the relationship between the load of nitrogen and the level of nitrate can be described by a linear function. The production function was therefore given a linear form.

The estimated nitrogen purification capacity of wetlands was based on the results of a Swedish study, according to which the denitrification of mature wetlands varies between 100 and 500 kg N/ha/year depending on type of wetland and on the locality. However, the maximum abatement capacity of restored wetlands is only achieved after about 3–5 years (see e.g., Kusler and Kentula 1990). In order to account for this delay a relatively low level of nitrogen purification was assumed, 200 kg N/ha/year.

In order to estimate the value function of the ancillary benefits of restoring wetlands the results from a study of the life-supporting values of a mire in Gotland were used; see Section 4.6. Several life support functions were included in the study such as nitrogen abatement, water buffering, supply of energy and provision of habitat. These life-support functions were evaluated at their replacement costs.

Excluding the value of nitrogen abatement, the total value of the life-support functions was estimated to range between SEK 9600–25000/ha/year. However, since this value was estimated for a mature wetland it was assumed that the corresponding value for a restored wetland would be

Table 4.4 *Nitrogen sources and their emissions (tons per year)*

Source	Emission
Agriculture	8,510
Sewage treatment plants	170
Air deposition	
• Sources in Gotland	775
• Imports	2,325
Total	11,780

Source: See text.

Table 4.5 *Marginal values of nitrogen abatement (Swedish kroner per kilogram)*

Source	Water quality	Ancillary benefits	Total
Restoration of wetlands	259	600	859
Sewage treatment plants	104		104
Agriculture	5		5

Source: See text.

half of the lower bound value, i.e., SEK 4800/ha/year. It was further assumed that the production function for secondary benefits was linear in the stock of wetlands.

Restoration of wetlands is a recently established area of research and few experimental results are available which could be used in this model. It was therefore simply assumed that the rate of natural growth in the stock of wetlands is constant and amounts to .01/year. In order to compare the values of a marginal investment in sewage treatment plants and in wetlands it was further assumed that the real discount rate, r, is 0.03 and that the economic life of a sewage treatment plant is 50 years, which implies that the rate of depreciation, P, is 0.02.

Given all these assumptions, the total value of a marginal increase in the restoration of wetlands, in the capacity at the sewage treatment plant, and for a marginal decrease in farmers' use of nitrogen, are presented (in SEK/kg nitrogen abatement) in Tables 4.4 and 4.5.

According to the results presented in Table 4.4, the value of a marginal increase in nitrogen abatement by wetlands is considerably higher than the corresponding values of the alternative methods. This is partly explained by the value of the ancillary benefits which account for about two-thirds of the total marginal value of restoring wetlands. Another reason is the differences in the rates of growth in the capacity to produce services. This can be seen in the case of the value of water quality where the results are significantly affected by the differences in the rates of growth.

In other words, investing in ecosystem restoration (i.e., restoration of the primary value source) in order to generate a nitrogen abatement (a component secondary value) also generates other valuable environmental services (other component secondary values).

4.9   Discussion

Many of the wetland functions and services discussed in this chapter do not have a direct market value. This is one fundamental reason why the wetlands' often unperceived but real and long-lasting societal support values (total primary value) have been destroyed or degraded via conversion to land use activities that generate a short-term, directly capturable and immediate income stream.

The aggregate value (total secondary value) of the environment as a factor of production, and the free support of ecosystems (including the "glue" value) are not yet fully recognised within economic systems, although economies are dependent on this support to be able to function (Odum 1989). For example, the human-made technical replacements discussed in the biophysical case study above, were installed because of a need in society to mitigate environmental and natural resource degradation and loss problems. But when the replacements were in place it was generally forgotten that the original need for the replacement investment was the loss of the wetland's already existing life-support functions. The TEV concept (based on willingness to pay), used by many economists for environmental valuation, is not able to fully capture the value of the fundamental role of "healthy" ecosystems in sustaining socioeconomic activities (the primary value of ecosystems) which at a basic level underpins human economies. At present it deals mainly with measuring values of output, or the secondary values provided by ecosystems, and then only with some of them (often not the life-supporting component functions of the ecosystem). It is difficult to capture the full significance of the life-support functions through human based preferences valuation alone, our "glue" value is a nonpreference-based instrumental value.

The study on wetland restoration for nitrogen purification provides a promising way to include more of this instrumental value, and such studies could contribute to make human preferences more aware of the benefits derived from life-support systems. The study revealed that restoring wetlands (the primary value) to gain one ecological services (nitrogen abatement) also resulted in gaining several other secondary environmental values. This is due to the multifunctionality of the wetland, which resulted in large benefits relative to more conventional technologies. The "extra" benefits accounted for about two-thirds of the total marginal value of restoring wetlands, which was about 8.5 times larger than for sewage treatment plants (see Table 4.5). Restoring wetlands also creates "new" habitats for species, and generally makes the landscape more diverse. In fact, using a living system in this way implies using biological diversity for the production of goods and services.

On balance, given the historical and on-going loss of wetlands and the argument that most wetlands, once destroyed, can only be partially and imperfectly replaced by man, a precautionary approach to further wetland exploitation is strongly recommended. In the context of project appraisal involving development versus wetland conservation conflicts, it would seem appropriate to require that cost-benefit analysis be used to choose between alternatives only within a choice set bounded by sustainability (ecosystem stability and resilience) constraints (Common and Perrings, 1992; Nash and Bowers, 1988). All this is not, however, to argue that economic valuation of wetlands is irrelevant. Our case studies have demonstrated that many of the wetland functions and services can be meaningfully valued in monetary terms. Although such analysis can only be partial, and therefore will probably yield an underestimate of the "true" total value of wetlands, it can still play an important part in the heuristic cost-benefit analysis. For instance, estimates of only a few secondary values may swing the balance in favour of wetland conservation in comparison to alternative use benefits. The study of Broadland is a good example of this (Table 4.3, Section 4.7), where it was shown that some secondary values have strong economic multiplier effects for local areas or regions.

One further caveat that should always be borne in mind in wetland ecosystem valuation is that the component functions and services of the system are often interrelated and sometimes mutually exclusive. Simple aggregation of component values is not a valid general rule. Recognition of the interdependence and hierarchy of values as discussed in this chapter is crucial if the resource base underpinning human societies is to be sustainably managed.

CHAPTER 5

An ecological economy: notes on harvest and growth

Gardner Brown and Jonathan Roughgarden

5.1 Introduction

These notes offer an interdisciplinary contribution to economics and ecology. The focus is on the coexistence of economic and ecological prosperity. We discuss two topics. We show how best to harvest a resource such as a marine fishery based on a contemporary metapopulation model from ecology, and we also offer a replacement for the neoclassical model of economics that shows how an economy depends on the aggregate of its natural resources, the human population size, and the investment policy. The work on these two topics has been carried out in part simply to see how ecological and economic theory may be combined. The theme of biodiversity is also involved, however, because it emerges that the optimal harvest of a fishery with a pelagic larval phase depends on the biological diversity, not of species, but of habitats in which the adults live. Also, in the macroeconomic realm, work on the second topic shows a dependence of economic prosperity on the stock of natural resources, and implies the need to conserve natural resources from extinction if economic prosperity is to be attained.

From an ecological perspective, policies for the development of natural resources should rest on the more realistic models of population dynamics that have emerged in recent years from ecological research, rather than on early ecological textbook models such as the logistic equation. An equation such as the logistic remains valuable when working through practice exercises, just as myths like ideal pulleys and gases remain valuable in teaching mechanics and chemistry. These notes show how to harvest in the most economically profitable way a population that has a two-phase life cycle, such as occurs in most coastal marine organisms. The larval stage forms a "larval pool" that is, in effect, a "commons" that serves recruits to all the local sites where the adults live.[1] The dynamics of the population is stabilised by competition among adults for space within the local sites. No density dependence is present in the larval phase. This type of model is an example of a "metapopulation" model in ecology, and represents a spatially structured population (Possingham and Roughgarden 1990; Roughgarden et al. 1985; Roughgarden and Iwasa 1986). Harvesting takes place at the local sites, where the adults are assumed to be the organisms that are edible. It is discovered that optimal harvesting should specialise on no more than one of the sites, while the others should be conserved as "nurseries."

This population model shares few properties with others in the fishery economics literature. Unlike predator-prey and related multiple species models, the adult populations do not compete directly with each other. There is a slight resemblance to cohort fishery models treated by economists (Conrad and Clark 1987) in as much as surviving larvae become adults but the common larval pool adds distinctive structure and leads to strikingly different conclusions that would hold even if the life stages were dated in explicit age classes.[2]

Also from an ecological perspective, the view of economic growth embodied in the fundamental "neoclassical model" of the economy is impossible. In the classic view, an economy is viewed as having two state variables, capital and labour. Production (GNP) is assumed to be a function solely of these two state variables, with the environment remaining a passive bystander. During the conventional analysis of the neoclassical growth model, the equations for these two state variables are collapsed into one equation by dividing through by the size of the labour pool, yielding a simpler model in one state variable that is based on per-capita quantities. When distilling the state variables of capital and labour to one per-capita variable, two assumptions are explicitly used. First, the labour pool is assumed to be growing exponentially at some externally fixed rate. Indefinite positive exponential population growth is impossible. Second, the relation between production, capital and labour is assumed to possess "constant returns to scale." This means that if the number of workers is doubled, and the amount of capital is also doubled (so that the capital per worker remains the same), then production also doubles. This too is

---

[1] It is a common pool in the sense of Bromley (1991) because there are biologically explicit rules for allocating the pool across "demanders." It is not an open access commons in the sense of Hardin's (1968) Tragedy of the Commons.

[2] For example, in one treatment of multiple cohorts, recruitment is independent of stock, and the entire population is pulse harvested sequentially, a solution which resembles the optimal harvest plan for a simple forest (Clark 1990). In this model recruitment does functionally depend on stock, and the optimal harvest policy is quite different.

impossible. A real economy depends on natural resources such as food and space, and the use of these resources causes a dynamic response of the natural ecosystem that supplies these resources. No real ecosystem responds to harvesting in a way that preserves constant return to the scale of exploitation.

Here, a replacement for the neoclassical model of the economy is offered. This replacement is an *ecological economy*. It has three state variables, labour and capital as before, and a third variable, the amount of natural resource in the environment. The economy's production comes from harvesting this resource. The harvesting *effort* is assumed to have constant returns to scale (two little boats work as hard as one big boat) but the *yield* from that effort, which is the economy's production, does not possess constant returns to scale because the yield depends on the state of the natural resource. Next, the labour force is assumed to have its own dynamic that depends upon an allocation from some of the economy's production – the growth of labour is an internal process dependent on harvest. There is one control variable. Central planning, assumed wise and well-meaning, controls the investment of total production to capital and to consumption. Moreover, there is a parameter, that could also be taken as a control variable in future studies, that indicates how direct personal preference controls the allocation of personal consumption to producing more labour (having children) and to making discretionary expenditures (buying consumer goods). Optimal policies may maximise any of several criteria, discounted cumulative per-capita consumption, labour size, or size of the natural resource stock, depending on whether the model is intended to be descriptive or normative.

5.2   Optimal harvesting: simple closed population

We begin by reviewing some well-known theory for optimally exploiting a single closed density-regulated population. The dynamics of the population being harvested is

$$\dot{N} = F(N) - h \qquad (5\text{-}1)$$

and the objective is to find the harvest rate that maximises wealth, defined as the cumulative harvest discounted by rate r

$$\max \int_{0}^{\infty} e^{-\rho t} P h \, dt = W \qquad (5\text{-}2)$$

where P is the market price per harvested individual. The current value Hamiltonian is

$$H = Ph + \lambda[F(N) - h] \tag{5-3}$$

where λ is the shadow price that can be viewed as what the harvester must pay the ecosystem for each organism harvested. Then, if $h > 0$,

$$\frac{\partial H}{\partial h} = P - \lambda = 0 \tag{5-4}$$

implies that

$$\lambda = P \tag{5-5}$$

Next,

$$-\frac{\partial H}{\partial N} = \dot{\lambda} - \rho\lambda = -\lambda F'(N) \tag{5-6}$$

implies that

$$\rho = F'(N) \tag{5-7}$$

because $\dot{\lambda} = 0$. In summary then, optimal harvesting is brought about by catching organisms at rate h where

$$h = F(N) \tag{5-8}$$

and where N is the root of

$$\rho = F'(N) \tag{5-9}$$

and the wealth thereby accumulated, once steady state is achieved, is

$$W = \int_0^\infty e^{-\rho t} P h \, dt = \frac{Ph}{\rho} \tag{5-10}$$

Logistic resource

As an exercise, we briefly state the textbook example of exploiting a logistic population,

$$F(N) = \frac{r N (K - N)}{K} \tag{5-11}$$

where r is the intrinsic rate of increase and K is the carrying capacity. This function is a quadratic opening downwards and intersecting the horizontal axis at 0 and K, with a peak at $K/2$. The maximum biologically sustainable yield is obtained when the stock size, N, equals $K/2$. The stock size, N, at the economically optimal harvest is found from

$$F'(N) = \frac{r (K - 2N)}{K} = \rho \qquad (5\text{-}12)$$

yielding

$$N = K \phi \qquad (5\text{-}13)$$

and an optimal harvest rate, h, of

$$h = r K \phi(1 - \phi) \qquad (5\text{-}14)$$

where ϕ is

$$\phi = \frac{1}{2}\left(1 - \frac{\rho}{r}\right) \qquad (5\text{-}15)$$

The quantity ϕ is in $[0,1/2]$. Graphically, the economically optimal harvest is obtained when the slope of $F(N)$ equals the economic discount rate, r, a condition that lies to the left of the peak of $F(N)$. The economically optimal harvest equals the maximum biologically sustainable yield only if the economic discount rate r is 0. To develop the environment in a way that lowers its K for the harvested resource, the developer should pay the harvester an amount equal to

$$\frac{dW}{dK} = \frac{r\phi(1 - \phi)}{\rho}P \qquad (5\text{-}16)$$

If r is less than ρ the optimal economic policy is to harvest the population to extinction; if $r > \rho$ the economically optimal policy leads to a steady state in which a sustainable profit is being realised and the population continues to exist.

5.3   Barnacle model

The "barnacle model" is a population model that represents a two-stage life-cycle – larva and adult. The adults are associated with areas; they attach to rocks or other surfaces. The limiting resource for adults is space to live in. The larvae live in the ocean's waters, and form a "larval pool." Eggs produced by the adults hatch to larvae that enter the ocean. After a development period, the larvae settle out of the water onto surfaces and metamorphose into adults. This model can represent abstractly most coastal marine populations, including commercially harvested stocks such as fish, lobsters and shrimp, and mussels, because most coastal marine organisms have a two phase life-cycle qualitatively similar to that of barnacles.

Single-stock with internal larval pool

The dynamics of the adult stock, N are given by

$$\dot{N} = cL\,(A - aN) - \mu N \tag{5-17}$$

where the first term is the input – settlement onto vacant space – and the second term is the output – mortality. A is the amount of space in the habitat, a is the area occupied by an individual, so aN is the amount of space occupied by the organisms. $A - aN$ is the vacant space. The number of larvae is L and these settle into this vacant space with rate constant c. m is the mortality rate of adults. The dynamics of the larval stock, L, are given by

$$\dot{L} = mN - cL\,(A - aN) - vL \tag{5-18}$$

The first term is the input – reproduction by adults with fecundity rate m, the second term is the loss due to settlement which is exactly what was the input term in the preceding equation, and the third term is the mortality to larvae in their oceanic habitat at rate n.

To begin, one may imagine that the larval pool instantaneously adjusts or "tracks" the state of the adult stock. If so, we can view the larval stock as always being in equilibrium with the current N. By setting $\dot{L} = 0$, and solving for L, as a function of N we have

$$L = \frac{mN}{c(a-aN) + v} \tag{5-19}$$

Substituting this L into the equation for $\dot{N}$ yields a population dynamic model for a stock that we can harvest at rate h,

$$\dot{N} = F(N) - h = \left(\frac{cm(A - aN)}{c(A - aN) + v} - \mu \right) N - h \tag{5-20}$$

The graph of $F(N)$ is $\cap$-shaped and intersects the horizontal axis at 0 and at

$$N_m = \frac{A}{a} - \frac{\mu v}{ac(m - \mu)} \tag{5-21}$$

The derivative of $F(N)$ is

$$F'(N) = -\mu + \frac{cm(A - an)}{c(A - aN) + v} - \frac{acmvN}{(c(A - an) + v)^2} \tag{5-22}$$

The stock size at the optimal harvest works out to be one of two roots of a quadratic equation. The roots are denoted with the symbol $\pm$ to indicate both the root where a plus sign is used, and that where a minus sign is used. Only one of these roots is relevant, the other should be discarded (which is apparent numerically). The roots are

$$N_{\pm} = \frac{A}{a} + \frac{v}{(a\,c)} \pm Z \tag{5-23}$$

where

$$Z = \left(\frac{mv(Ac + v)}{a^2 c^2 (m - (\mu + \rho))} \right)^{1/2} \tag{5-24}$$

Therefore the optimal harvest is

$$h_{\pm} = \frac{m - \mu \pm mv}{(a\,c\,Z)} N_{\pm} \tag{5-25}$$

The steady state wealth obtained from this harvest is

$$W_{\pm} = \frac{P\,h_{\pm}}{\rho} \tag{5-26}$$

A developer who wishes to buy some of the area for other uses should pay the harvester

$$\frac{dW_{\pm}}{dA} = \frac{P}{\rho} \frac{dh_{\pm}}{dA} \tag{5-27}$$

where

$$\frac{dh_{\pm}}{dA} = - \frac{m^2 v^2 (\pm(Ac + v)) + acZ}{2a^4 c^3 (m - (\mu + \rho))Z^3}$$
$$+ \left(m - \mu \pm \frac{mv}{acZ} \right) \left(\frac{1}{a} \pm \frac{cZ}{2(Ac + v)} \right) \tag{5-28}$$

External larval pool

In this section we do not assume that the larval pool instantaneously tracks the adults, that is, the larval pool is external not internal. A model for a metapopulation of barnacles with an external larval pool being harvested at the adult sites, $i = 1 \ldots w$ is

$$\dot{N}_i = F_i(N_i, L) - h_i \tag{5-29}$$

$$\dot{L} = G\,(N_1 \ldots N_\omega, L) \tag{5-30}$$

where

$$F_i(N_i) = c_i\,L\,(A_i - a_i\,N_i) - \mu\,N_i \tag{5-31}$$

$$G(N_1 \ldots N_\omega, L) = \sum_{i=1}^{\omega} m_i N_i - \sum_{i=1}^{\omega} c_i\,L\,(A_i - a_i N_i) - vL \tag{5-32}$$

The objective is

$$\max \int_0^\infty e^{-\rho t} \left(\sum_{i=1}^{\omega} P_i h_i \right) dt \tag{5-33}$$

The current value Hamiltonian for this problem is

$$H = \sum_{i=1}^{\omega} P_i h_i + \sum_{i=1}^{\omega} \lambda_i (F_i - h_i) + \lambda_L\,G \tag{5-34}$$

Maximising H with respect to the control variables leads, for each i, and provided $h_i > 0$ to

$$\frac{\partial H}{\partial h_i} = P_i - \lambda_i = 0 \tag{5-35}$$

so that, for each i being harvested, the shadow price is

$$\lambda_i = P_i \tag{5-36}$$

Next, for each i, the rate of change of shadow price is

$$-\frac{\partial H}{\partial N_i} = \dot{\lambda}_i - \rho\lambda_i = -\left(\lambda_i \frac{\partial F_i}{\partial N_i} + \lambda_L \frac{\partial G}{\partial N_i} \right) \tag{5-37}$$

At equilibrium, $\dot{\lambda}_i$ is 0, so, for each i at which harvesting takes place

$$\lambda_L = P_i \frac{\rho - \dfrac{\partial F_i}{\partial N_i}}{\dfrac{\partial G}{\partial N_i}} \tag{5-38}$$

Also, another expression for λ_L comes from

$$-\frac{\partial H}{\partial L} = \dot{\lambda}_L - \rho\lambda_L = -\left(\sum_{i=1}^{\omega} \lambda_i \frac{\partial F_i}{\partial L} + \lambda_L \frac{\partial G}{\partial L} \right) \tag{5-39}$$

which, at equilibrium, implies

$$\lambda_L = \frac{\sum_{i=1}^{\omega} P_i \frac{\partial F_i}{\partial L}}{\rho - \frac{\partial G}{\partial L}} \qquad (5\text{-}40)$$

A further equation pertaining to the equilibrium values of N_i and L is

$$G(N_i \dots N_w, L) = 0 \qquad (5\text{-}41)$$

Finally, harvest rates corresponding to the optimal stocks are found, for each i, from

$$h_i = F_i(N_i, L) \qquad (5\text{-}42)$$

One adult stock

In the case of one adult habitat, $\omega = 1$, the stocks of adults and of larvae at equilibrium are found from

$$\frac{\rho - \frac{\partial F}{\partial N}}{\frac{\partial G}{\partial N}} = \frac{\frac{\partial F}{\partial L}}{\rho - \frac{\partial G}{\partial L}} \qquad (5\text{-}43)$$

$$G = 0 \qquad (5\text{-}44)$$

The optimal harvest rate that leads to the optimum stock sizes is then

$$h = F(N, L) \qquad (5\text{-}45)$$

Explicitly, the equations for determining N and L in this model are

$$\frac{acL + \mu + \rho}{acl + M} - \frac{c(A - aN)}{c(A - aN) + v + \rho} = 0 \qquad (5\text{-}46)$$

$$mN - cL(A - aN) - vL = 0 \qquad (5\text{-}47)$$

The optimal steady state values of N and L depend on the discount rate but are independent of other economic parameters such as the net value of harvest, P. It holds when there are one or more demand functions and when there is dependence among the demand functions. As before, the price to be paid by a developer to the harvester for diverting area, A to other purposes can be obtained by differentiating the steady state wealth accumulated with this harvesting policy with respect to A. Also, the price for interfering with the larval pool, and killing a larva, is given by λ_L, which, incidentally, is less than λ_N.

Many adult stocks

The case of many adults stocks in this model has a surprising equilibrium introduced by writing out Equation (5-39) explicitly

$$\rho = a_i c_i L \left(\frac{\lambda_L}{\lambda_i} - 1\right) + \frac{\lambda_L}{\lambda_i} \qquad i = 1 \ldots \omega \tag{5-48}$$

Assume naively, interior solutions throughout, setting $\lambda_i = P_i$, then the ω equations in $\omega + 1$ unknowns of the equations above collapse to

$$\rho = \frac{a_i c_i L}{P_i} (\lambda_L - P_i) + \frac{\lambda_L}{\lambda_i} \qquad i = 1 \ldots \omega \tag{5-49}$$

Here there are ω equations in two unknowns, L and λ_L. Except fortuitously, this exercise demonstrates that it is optimal to harvest from at most only two spatial populations. In fact it is optimal to harvest from only one site. To see this, assume two sites with the same product price, which differ only in the productivity of the N_i in the production of L; i.e., $m_1 \neq m_2$. The steady state conditions of Equation (5-39) can be rewritten as

$$\lambda_1 = \lambda_L \frac{acL + m_1}{acl + \rho + \mu} \tag{5-50}$$

$$\lambda_2 = \lambda_L \frac{acL + m_2}{acl + \rho + \mu} \tag{5-51}$$

Assuming no economic specialisation so there is harvest at both sites, $\lambda_1 = P = \lambda_2$, then the equations above can be solved if the biological parameters of the model, m_1 and m_2, are equal. In general this condition does not hold. The site most productive in producing L, say, site 2, $m_2 > m_1$, should specialise in that activity.

The economic motivation for such extreme specialisation, where harvest occurs at only one site, independent of the number of sites, is explained by the biological productivity of L. In the steady state,

$$L = \frac{\sum m_i N_i}{\sum c_i F_i + v} \qquad i \neq 1 \tag{5-52}$$

Thus, the marginal product of N_i in the production of L is positive,

$$\frac{\partial L}{\partial N_i} = \frac{m_i}{c \sum F_i} + \frac{a_i c \sum m_i N_i}{(c \sum F_i)^2} > 0 \tag{5-53}$$

which is natural enough, but also

$$\frac{\partial^2}{\partial N_i^2} = \frac{2a_i^2 \Sigma m_i N_i}{(c\Sigma F_i)^3} > 0 \tag{5-54}$$

So the marginal product of N_i in the production of L is increasing, a condition of increasing marginal returns rarely found in economic models but naturally occurring in this biological structure. If some N_i is good at producing L, more is better. Any harvest from a site producing larvae detracts from enhancing L, which acts as a multiplier in the formula for the optimal harvest.

Another way to understand the solution of extreme specialisation in the model is to think of the ω local adult sites as countries engaged in trade. All the countries containing a stock contribute to the production of L and sell it at a price λ_L. The country with the comparative advantage in harvest, produces harvest. It maximises harvest by having no adult stock of its own in each period. Any positive stock detracts from harvest. This striking result is made possible by the biological common pool, a technical externality that permits a continuous infusion of harvestable biomass to each site. Unfortunately though, there is no transparent approach to choosing the optimal harvest site because it depends on all the physical and economic parameters in the model.

5.3   Prosperity in an ecological economy: a macro ecological-economic model

In the fundamental neoclassical model of economics, two state variables are initially introduced, K, the aggregate capital in the economy, and L, the aggregate labour force. By assuming exponential growth of the labour population according to some rate fixed determined outside the economy, and by assuming the GNP possesses a constant return to scale as a function of K and L, these two state variables are collapsed down to one, the per-capita capital, k, which is by definition, K/L. In the one-variable model, optimal growth is defined as the per-capita consumption, c, that maximises the discounted cumulative c – by investing the economy can grow leading to more consumption later on, and the optimal c strikes the best balance between future economic growth and present-day consumption. Unfortunately, the mathematical steps involved in converting the two-variable problem into a one-variable problem assume an impossible and infinite world, exponential growth of the labour force and a production function that has constant returns to scale. Therefore, we offer a model to replace the neoclassical model, a model we term an "ecological economy."

The basis of production in an ecological economy is a harvestable natural resource whose stock is N, and whose dynamics are

$$\dot{N} = F(N) - a_N H(K,L) N \qquad (5\text{-}55)$$

where a_N is a coefficient that ensures proper units. $F(N)$ is the natural dynamics of the resource, and can represent exponential, logistic, or other population-dynamic models as discussed above in connection with optimal harvesting. Also, F might be written as $F(N,K,L)$ to model a positive influence of capital (or technology) on improving the growth rate of the resource, and to model a negative influence of loss of habitat, pollution and human population size on reducing the growth rate of the resource. The total harvest is $H(K,L)N$ where H is the harvesting effort. This effort is a function of the capital in the economy, K, and the number of workers, L. We assume $H(K,L)$, the harvesting effort involves constant return to scale. Typically, we take

$$H(K,L) = K^\alpha L^{1-\alpha} \qquad (5\text{-}56)$$

where α determines the scale whereby capital and labour are substitutable. (For simplicity, we conventionally take $\alpha = 1/2$, so that K and L have the same elasticity.) The yield from the harvesting effort depends on the size of the stock being harvested, and does not show constant return to scale because the yield is directly proportional to N, the stock size. This functional dependence resembles predator-prey models in ecology where the harvesting rate by a predator population, P, of a prey population, V, is normally assumed to be proportional to PV. An alternative assumption is that the total harvest (not solely the harvesting effort) is a concave function such as $a_h K^\alpha L^\beta N^\gamma$ where $\alpha + \beta + \gamma = 1$. This alternative would weaken the ability of the economy to control the resource, just as predator satiation allows prey outbreaks in ecological predator-prey models. If this alternative is preferred, then any stability in the economy would be attributed to significant density dependence within the resource's dynamics, $F(N)$, and not to the ability of the economy to absorb the resource's productivity. Thus, for the present, the formulation above is preferred, but can be revised if desired.

The production by the economy (GNP) is then allocated, presumably by a wise and well-meaning central plan, into investment and consumption, where C, the total consumption, is a control variable

$$a_K H(K,L) N = \dot{K} + \mu K + C \qquad (5\text{-}57)$$

a_K ensures the correct units. Investment goes specifically into increasing the amount of capital, $\dot{K}$, and to restoring depreciated capital μK.

Consumption, C, is further allocated, presumably by personal preference, into reproduction and discretionary consumption, where D, the total discretionary consumption, is a parameter

$$C = \left(\frac{1}{a_L}\right)(\dot{L} + v\, L) + D \tag{5-58}$$

The a_L ensures correct units. Reproduction goes specifically into growth of the labour force, $\dot{L}$, and to replacing mortality, vL. Discretionary income is used to purchase consumer goods. Also, the mortality rate, v, can be written as $v(K,L)$ to indicate an effect of capital (and technology) on lowering mortality, and of human population size in increasing mortality from social stress, higher incidences of parasitism and disease, and loss of ecosystem services (Ehrlich and Ehrlich 1992).

Combining these assumptions leads to the following dynamical system

$$\dot{N} = F(N) - a_N\, H(K,L)\, N \tag{5-59}$$

$$\dot{K} = a_K\, H(K,L)\, N - \mu K - C \tag{5-60}$$

$$\dot{L} = a_L\, (C - D) - vL \tag{5-61}$$

where

$$0 \le C(t) \le a_K\, H(K,L)\, N \equiv \text{GNP} \tag{5-62}$$

$$0 \le D(t) \le C(t) \tag{5-63}$$

There are several possible choices of what to maximise when finding the optimal C. One choice is to maximise, as in the traditional neoclassical model of economics, the discounted cumulative per-capita consumption

$$\max \int_0^\infty e^{-\rho t}\, U\!\left(\frac{C(t)}{L(t)}\right) dt \tag{5-64}$$

where U is a concave utility function, say of the form

$$U(x) = x^\beta \tag{5-65}$$

with $\beta < 1$. This choice is plausible from the point of view of a central planner that controls the allocation of production into capital investment vs. personal consumption. Other value systems are to maximise the total number of people that have ever lived

$$\max \int_0^\infty L(t)\, dt \tag{5-66}$$

to maximise the total number of resource organisms that have ever lived

$$\max \int_0^\infty N(t) \, dt \tag{5-67}$$

or even to maximise the total number of all organisms that have ever lived

$$\max \int_0^\infty (L(t) + N(t)) \, dt \tag{5-68}$$

For discussion of the role of (human) population size in the objective functions, see Dasgupta (1982).

In these notes we consider the problem of an optimal ecological economy as one of choosing a single control variable, C, that determines how much of the economy's production is invested in capital and how much is diverted into consumption. However, in future studies, D could also be considered a control variable reflecting how individuals set their personal preferences. This would need to involve a submodel relating discretionary income to reproduction, and involve some form of the demographic transition. If D is also a control variable then the problem of an optimal ecological economy becomes a differential game, with the central planner controlling C and the individuals controlling, D. Both cooperative and noncooperative solutions could then be sought.

A similar, and intriguing, possibility is to embed biological evolution within the ecological economy, and to investigate a game between a national investment policy that determines C and biological evolution within the labour population that determines D. This topic may be investigated by placing genetic variation for D within the equation for $\dot{L}$ and considering D as a trait subject to density-dependent natural selection (Roughgarden 1971, 1979). Thus, the model offered here for an ecological economy can be investigated from a sociobiological standpoint, as well as from a wholly social science standpoint.

5.4   Dynamics of an ecological economy

To illustrate the model for an ecological economy, we offer an example based on an exponentially growing natural resource. One can think of this as a best case scenario for economic development. The model then becomes

$$\dot{N} = r \, N - a_N \, (K \, L)^{1/2} \, N \tag{5-69}$$

$$\dot{K} = a_K \, (K \, L)^{1/2} \, N - \mu \, K - C \tag{5-70}$$

$$\dot{L} = a_L (C - D) - v L \qquad (5\text{-}71)$$

where

$$0 \le C \le a_K (K L)^{1/2} N \equiv \text{GNP} \qquad (5\text{-}72)$$

$$0 \le D \le C \qquad (5\text{-}73)$$

The dynamics of this model involves three limit points. First, in the absence of labour or capital, the resource tends to infinity. Second, in the absence of resource, the capital and labour tend to 0. Third is the equilibrium at which the economy and the resource may coexist. The equilibrium, for a fixed C and D is

$$N = \frac{a_N C}{r a_K} + \frac{\mu_r}{a_K a_N} \frac{v}{a_L(C - D)} \qquad (5\text{-}74)$$

$$K = \frac{r^2}{a_N^2} \frac{v}{a_L(C - D)} \qquad (5\text{-}75)$$

$$L = \frac{a_L(C - D)}{v} \qquad (5\text{-}76)$$

$$\text{GNP} = C + \frac{\mu v r^2}{a_L a_N^2 (C - D)} > C \qquad (5\text{-}77)$$

This solution exists (is positive) if $C > D$, reflecting a requirement that not all the consumption can be allocated to discretionary expenses because the labour pool would die out. Thus, production by the ecological economy, coming from its harvesting of a natural resource, can absorb the exponential growth of the natural resource, resulting in an equilibrium point. All the equilibrium values vary monotonically with the discretionary income, D. L decreases with D, and K, N and GNP increase with D. Concerning the total consumption, increasing C increases L, decreases K, and has a cup-shaped effect on both the resource level and gross national product. A special case of interest is where all the consumption is allocated to human reproduction, $D = 0$, leading, for a given C, to the maximum human population size, minimum stock of capital, minimum stock of natural resource, and minimum GNP.

$$N_0 = \frac{a_N C}{r a_K} + \frac{\mu r}{a_K a_N} \frac{v}{a_L C} \qquad (5\text{-}78)$$

$$K_0 = \frac{a^2}{a_N^2} \frac{v}{a_L C} \qquad (5\text{-}79)$$

$$L_0 = \frac{a_L C}{v} \tag{5-80}$$

$$\text{GNP}_0 = C + \frac{\mu v r^2}{a_L a_N^2 C} \tag{5-81}$$

The stability of this equilibrium depends on the eigenvalues of the Jacobian evaluated at equilibrium,

$$\begin{pmatrix} 0 & -\frac{a_N^2}{2r}LN & -\frac{a_N^2}{2r}KN \\ \frac{a_K r}{a_N} & \frac{a_K a_N}{2r}LN - \mu & \frac{a_K a_N}{2r}KN \\ 0 & 0 & -v' \end{pmatrix} \tag{5-82}$$

The eigenvalues are

$$-v \tag{5-83}$$

$$\frac{1}{4}\left(-\mu + \frac{a_N^2}{r^2}CL + \sqrt{\text{desc}} \right) \tag{5-84}$$

$$\frac{1}{4}\left(-\mu + \frac{a_N^2}{r^2}CL - \sqrt{\text{desc}} \right) \tag{5-85}$$

where

$$\text{desc} \equiv \frac{a_N^2}{r^2}CL\left(\frac{a_N^2}{r^2}CL - 2(\mu + 4r)\right) + \mu(\mu - 8r) \tag{5-86}$$

These eigenvalues indicate that the equilibrium at which an economy coexists with a natural resource can be stable. For stability, the eigenvalues should be negative. One eigenvalue is always negative. Trajectories from anywhere within the N, K, L space approach, at rate $-v$, a plane parallel to the N, K plane that intersects the L axis at the equilibrium value of L. Within this plane, the equilibrium point for N and K is stable depending on whether the remaining two eigenvalues are also negative (if real) or have negative real parts (if complex).

The stability and dynamics of the ecological economy depends on the human population size, as determined by the individual preferences for allocating total consumption to reproduction and to discretionary purchases. Two cases are evident.

1. Equilibrium is stable regardless of D. The equilibrium can be stable regardless of D (and therefore of the human population size) if it is stable with $D = 0$,

$$\mu > \frac{a_L a_N^2 c^2}{vr^2} \tag{5-87}$$

   This can only be true, however, if the total consumption, C, is sufficiently low, and the critical maximum consumption level can be found by solving for C in the formula above where the $>$ is replaced by an $=$. If the total consumption is greater than this critical value, population limitation will be needed.

2. Equilibrium is stable for sufficiently high D. If the previous condition is not satisfied, then stability occurs if

$$\mu > \frac{a_L a_N^2 C(C - D)}{vr^2} \tag{5-88}$$

   This is true, for a fixed C, if D is large enough. The critical minimum allocation to discretionary income can be found by solving for D in the formula above where the $>$ is replaced by $a =$. Thus, provided the human population can be kept low enough (D high enough), the ecological economy is always stable. This case subdivides into two subcases.

   (a) Stable focus only. If

$$8r > \mu \tag{5-89}$$

   then the stable equilibrium is a stable focus. That is, trajectories in the N, K-equilibrium plane wind in to the equilibrium point as a damped oscillation.

   (b) Stable focus and stable node. If the inequality above is reversed, then for D sufficiently large, the equilibrium is a stable node in which trajectories in the N, K-equilibrium plane approach the equilibrium point from characteristic directions, and do not circle around the equilibrium point. For slightly lower D, the equilibrium is a stable focus.

For both these subcases, if D is low enough, then the equilibrium is unstable. Thus, the very stability of an ecological economy depends critically on population policy.

When the human population is too high to permit economic and ecological coexistence in a stable equilibrium, we conjecture that what results is a condition of continual change and uncertainty. Pending further analysis, we conjecture that the transition from a stable focus to an unstable focus when D is varied from a high to a low value produces a Hopf bifurcation to a limit cycle, which in ecological predator-prey models is known to have a striking bust-boom appearance. Furthermore, models in this circumstance can generate chaotic trajectories if the coefficients, such as r, are periodic. A periodic r reflects the natural summer-winter seasonality in the natural resource's renewability, and this period can interact with the inherent period in the ecological economy's limit-cycle oscillation, leading to chaos, i.e., to trajectories influenced by a strange attractor.

The preference for discretionary income, D, may not be only, or even largely, a matter of policy. If the labour force were viewed as a "typical" biological population, and not as a human population, the expectation would be for natural selection to cause D to evolve to 0. Inspection of Equation (5-72) shows that genes that lower D will, ipso facto, increase in the labour population, leading to the biological evolution of a progressively lower D. Nonetheless, the existence of the demographic transition implies that there are tradeoffs involved in lowering D indefinitely, making the extreme of $D = 0$ unlikely. Perhaps the most critical subject for future research is to understand the demographic transition from the standpoint of life history theory in ecology.

5.4   Policy for optimal growth

The policy variable that controls growth in the ecological economy is C, the total consumption. To find the optimal value(s) of this variable that maximises some discounted quantity, $I(C,D,N,K,L)$, we form the current-value Hamiltonian

$$H = I + y_N[rN - a_N(KL)^{1/2}N] + y_K[a_K(KL)^{1/2}N - \mu K - C]$$
$$+ y_L[a_L(C - D) - vL] \qquad (5\text{-}90)$$

where y_N, y_K, and y_L are the shadow prices (co-state variables) for N, K, and L respectively. To illustrate an optimal ecological economy that is most comparable to the neoclassical model, we suppose that the central planner is maximising the discounted cumulative per-capita consumption, so that $I = C/L$.

At equilibrium, the equations for these shadow prices, $\partial H/\partial x - \rho y_x = 0$ for $x = N, K, L$ are

$$\frac{a_K r}{a_N} y_K - \rho y_N = 0 \qquad (5\text{-}91)$$

$$\left(\frac{a_K a_N}{2r}LN - (\mu + \rho)\right) y_K - \frac{a_N^2}{2r}LN y_N = 0 \qquad (5\text{-}92)$$

$$\frac{a_K a_N}{2r} K N y_K - \frac{a_N^2}{2r} K N y_N - (\nu + \rho) y_L = \frac{C}{L^2} \qquad (5\text{-}93)$$

Next, we have $\partial H/\partial C = 0$ which implies

$$y_K - a_L y_L = \frac{1}{L} \qquad (5\text{-}94)$$

One should then solve Equation (5-93) for y_L and substitute into Equation (5-94) which is then solved for y_N. When this y_N is used with Equation (5-93) and with Equation (5-92) two equations for y_K are obtained. By setting these two equations for y_K equal to each other, an equation is derived in which the only unknown is C.

In this way, the investment policy that maximises per-capita consumption is found to be the real root of the following cubic equation for C

$$C^3 - \alpha_1 D C^2 + (\alpha^2 D^2 + \alpha_1) C - \alpha_2 \alpha_1 D = 0 \qquad (5\text{-}95)$$

where

$$\alpha_1 = \frac{\nu r^2 (\mu(r + \rho) + 2\rho^2)}{a_L a_N^2 (r - \rho)} \qquad (5\text{-}96)$$

$$\alpha_2 = \frac{\nu + \rho}{\rho} \qquad (5\text{-}97)$$

$$\alpha_3 = \frac{\nu + 2\rho}{\rho} \qquad (5\text{-}98)$$

It is assumed that $r > \rho$, implying that the α's are positive. This cubic equation has a factor, resulting in a very simple formula for the optimal C

$$C = \left(\frac{\nu}{\rho} + 1 \right) D \qquad (5\text{-}99)$$

Interestingly, in the limit when $D \to 0$, the optimal C also tends to zero. Thus, when the human population is near maximal, the maximal discounted per-capita consumption is achieved for members of the economy when total consumption is minimal. This, happily, is consistent with the stability conditions showing that when D is 0, C must be sufficiently low if the equilibrium is to be stable. Therefore, if both biological evolution and personal preferences conspire to produce the highest human population possible, managing the economy for the best per-capita consumption will tend to compensate for this destabilising factor, and lead instead to a sustained, stable, and prosperous ecological economy.

5.5   Conclusions

The theory of optimal harvesting is developed for a metapopulation of space-limited subpopulations, as is appropriate to many commercially harvested marine species. The striking conclusion is that the most profitable harvesting strategy involves harvesting solely at one subpopulation, while conserving the other subpopulations as nurseries.

A replacement for the neoclassical growth model of macro economics is proposed that includes a natural resource as the basis for the economy's production. Also (human) population growth is featured as an endogenous process dependent on the economy's production. This model shows that an economy and ecology can prosperously coexist, provided the human population is sufficiently low, which can be attained by sufficient allocation of resources to discretionary consumption.

Appendix

The model is

$$\dot{N} = rN - a_N (K L)^{1/2} N \tag{5-100}$$

$$\dot{K} = a_K (K L)^{1/2} N - mK - C \tag{5-101}$$

$$\dot{L} = a_L (C - D) - nL \tag{5-102}$$

where

$$\text{GNP} \equiv a_K (K L)^{1/2} N \tag{5-103}$$

$$C(t) = \min (C_0, \text{GNP}) \tag{5-104}$$

$$D(t) = 0, C/1.2 \tag{5-105}$$

The parameters for the numerical illustration are:

Parameter	Value
r	$\ln(2)$
m	.1
n	.01
a_N	5×10^{-6}
a_K	$10^{-5} a_N$
a_L	10
r	.05

These values are obtained by considering a scenario in which the natural resource is a fish population that supports a human population whose industry is fishing and whose capital investment is for fishing boats. Time is measured in units of years and space in km. Both N and L are in units of individuals, fish and people respectively. K is in units of "elemental boats," which we will think of as costing \$$10^4$. The fish are assumed to have the potential for doubling each year, so the intrinsic rate of increase for the natural resource, r, is $\ln(2)$. While the resource has the potential for exponential growth, both the capital equipment and human population have the potential for exponential decrease. We assume the life span of a fishing boat is 10 years, so μ is .1, and we assume the life span of a person is 100 years, making v equal .01. The remaining coefficients in the system model pertain to how the state variables of N, K, and L are coupled to one another.

We imagine that the fish live in an area of $1{,}000 \times 1{,}000 = 10^6$ km². The fish are further assumed to occur to a depth of 100 m, so that the volume they occur in is $1{,}000 \times 1{,}000 \times 0.1 = 10^5$km³. Therefore, the number of fish in a km³ is $N/10^5$. Next, we imagine that one person in an elemental boat travels at a velocity of 1 km per hour dragging a net that is 10 m wide $\times$ 10 m deep. The cross-sectional area is therefore .01 $\times$ .01 $= 10^{-4}$ km², and each hour this cross section is pulled through 1 km of water, so that the volume of water harvested by the elemental boat is 10^{-4} km³ per hour. If we allow 5,000 work hours per year (ca. 12 hours $\times$ 365 days), then the volume of water harvested by the elemental boat during the year is $5{,}000 \times 10^{-4} = .5$ km³ per year. So the harvest of fish from one person with an elemental boat per year is $N/10^5$ fish per km³ $\times$ .5 km³ per year. The harvest is, by definition, $a_N (K L)^{1/2} N$, so with both K and L being 1, we can identify a_N as $1/10^5 \times 0.5 = 5 \times 10^{-6.}$

Now for the remaining coupling coefficients. For a_K we assume each fish is worth about .1\$. Therefore, a_K, which simply converts harvest into units of capital, is .1 \$ per fish $\times$ 1 boat per 10^4\$ = \$$10^{-5} \times a_N$. For a_L, we take the cost of raising a child to be \$$10^5$, i.e., the cost of 10 elemental boats, so a_L is 10. The final coefficient is the inflation rate r which we take as 5%.

The figures illustrate how various properties of the model depend on C and D: the equilibrium values of N, K, and L; the real part and discriminant for the complex eigenvalues at equilibrium; and trajectories in the 3D phase space of the state variables. The trajectories were computed using a fourth order Runge Kutta algorithm. The trajectories describe 500 years, using as initial conditions .5 and 1.5 the equilibrium values of the state variables. For C from 500 to 2000 the step size was 3 months, for C from 3,000 to 4,000 the step size was 1 month, and for C from 5,000 to 6,000 the step size was 1 week.

Two values of D were used, 0 and $C/1.2$. The later is the D for which the C is optimal with the values of v and ρ above.

In the figures notice that the stable equilibrium bifurcates to a limit

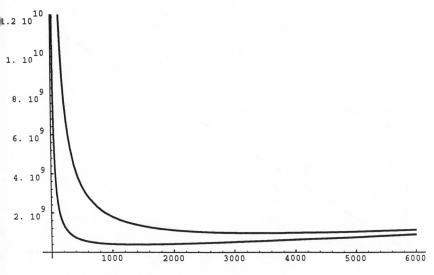

Figure 5.1. Neq as a function of C for $D = 0$, $C/1.2$.

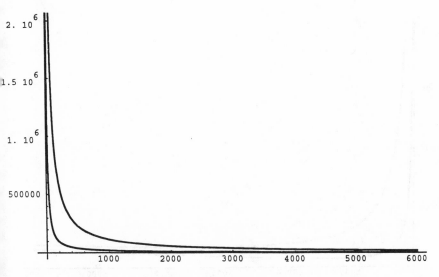

Figure 5.2. Keq as a function of C for $D = 0$, $C/1.2$.

cycle at the value of C where the real part of the complex eigenvalues is 0. Notice also that the sequence of illustrations where D is $C/1.2$ show a generally more stable ecological economy than the sequence where D is 0.

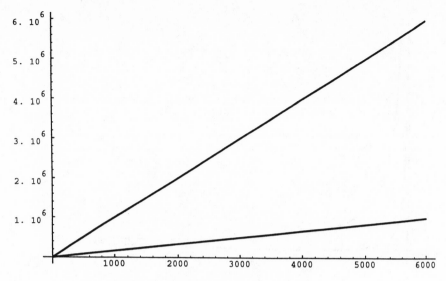

Figure 5.3. Leq as a function of C for $D = 0$, $C/1.2$.

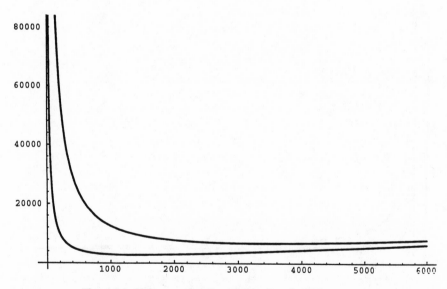

Figure 5.4. GNPeq as a function of C for $D = 0$, $C/1.2$.

172

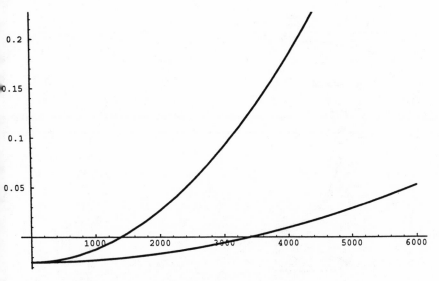

Figure 5.5. Re $l_{2',3}$ as a function of C for $D = 0$, $C/1.2$.

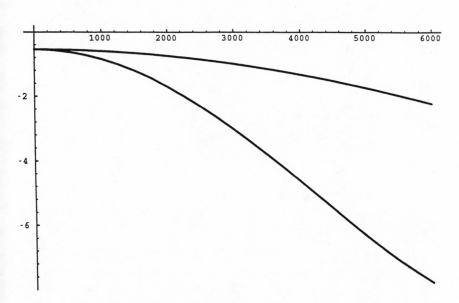

Figure 5.6. Descriminant as a function of C for $D = 0$, $C/1.2$.

173

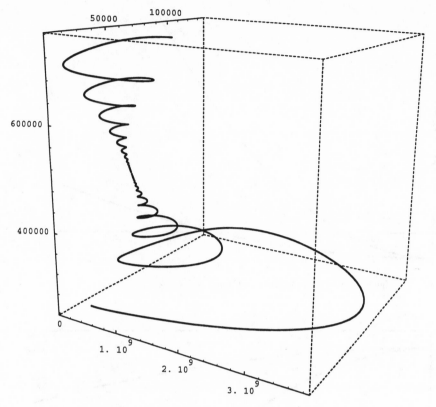

Figure 5.7. $C = 500$, $D = 0$.

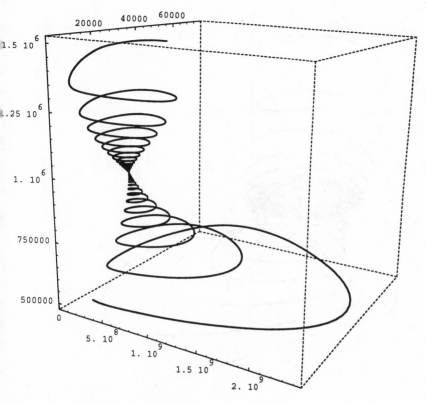

Figure 5.8. $C = 1,000$, $D = 0$.

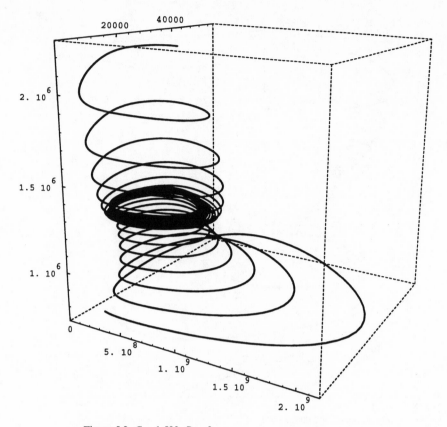

Figure 5.9. $C = 1,500$, $D = 0$.

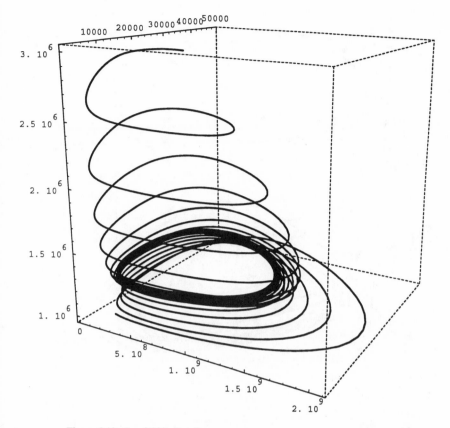

Figure 5.10. $C = 2,000$, $D = 0$.

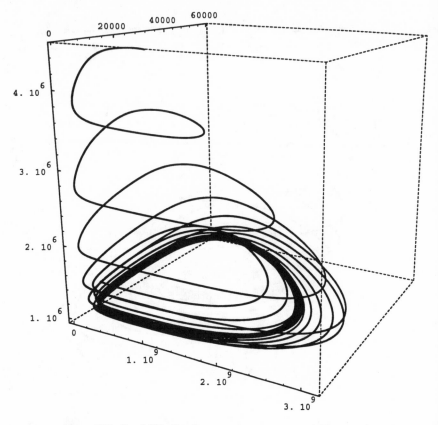

Figure 5.11. $C = 3,000$, $D = 0$.

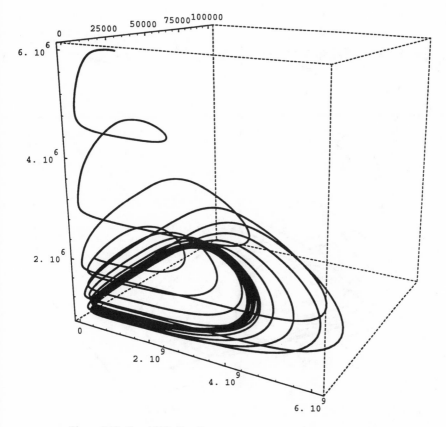

Figure 5.12. $C = 4,000$, $D = 0$.

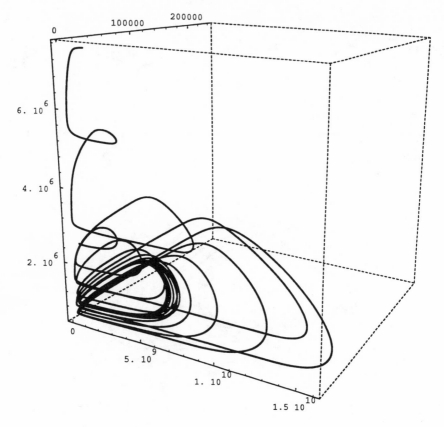

Figure 5.13. $C = 5,000$, $D = 0$.

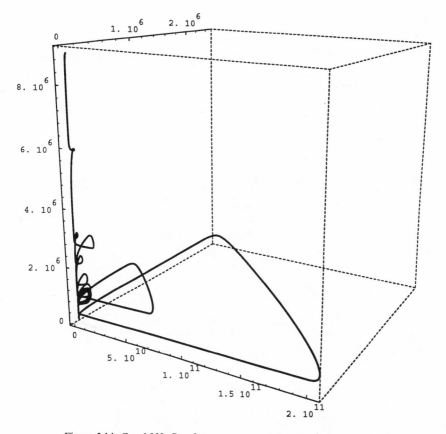

Figure 5.14. $C = 6{,}000$, $D = 0$.

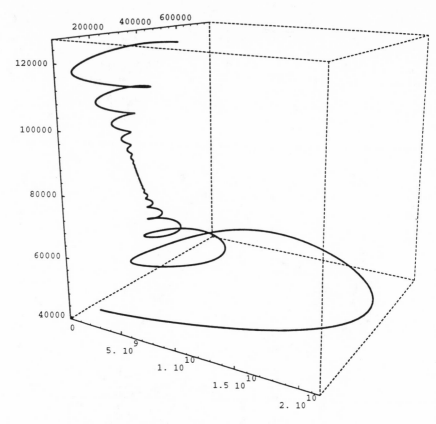

Figure 5.15. $C = 500$, $D = C/1.2$.

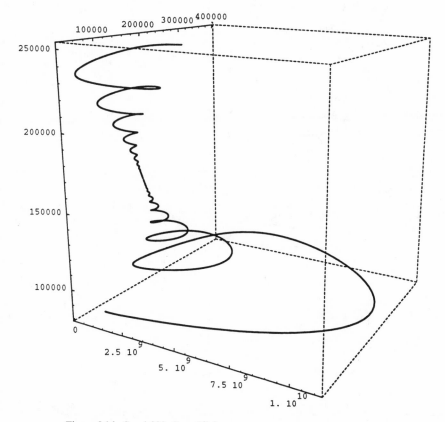

Figure 5.16. $C = 1,000$, $D = C/1.2$.

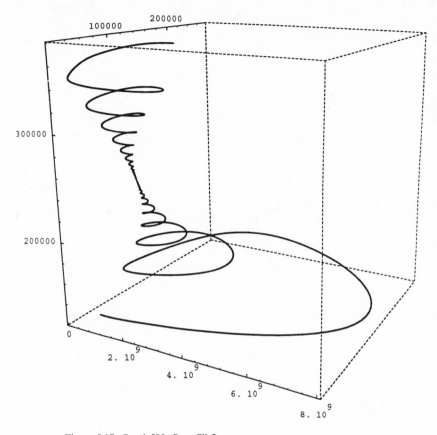

Figure 5.17. $C = 1,500$, $D = C/1.2$.

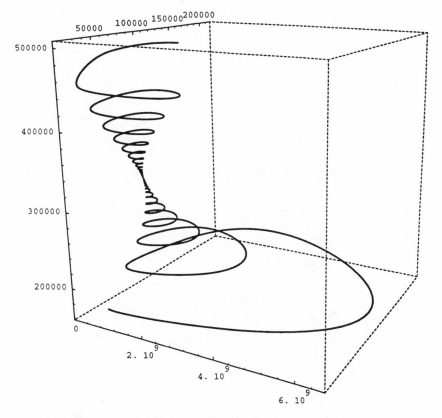

Figure 5.18. $C = 2,000$, $D = C/1.2$.

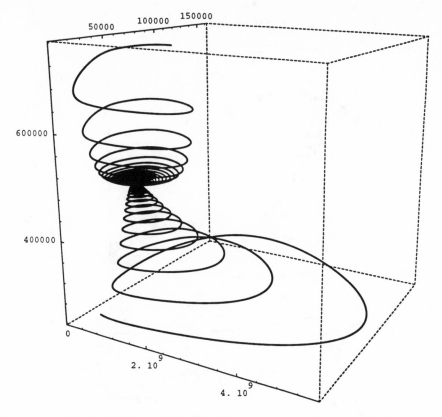

Figure 5.19. $C = 3,000$, $D = C/1.2$.

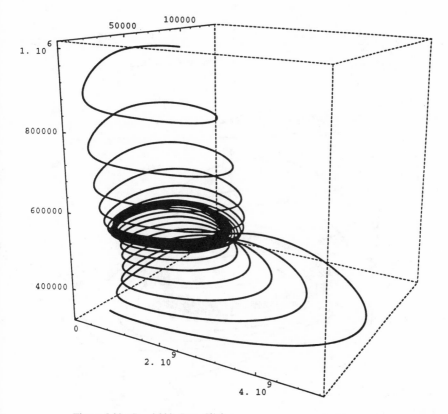

Figure 5.20. $C = 4,000$, $D = C/1.2$.

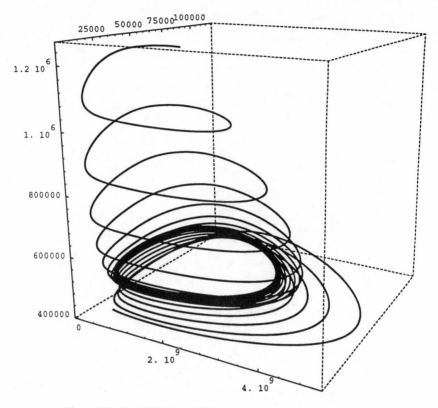

Figure 5.21. $C = 5{,}000$, $D = C/1.2$.

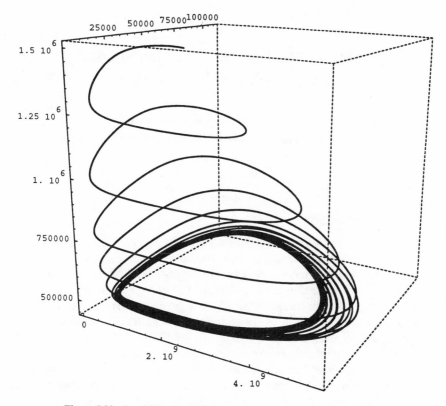

Figure 5.22. $C = 6,000$, $D = C/1.2$.

CHAPTER 6

Biodiversity loss and the economics of discontinuous change in semiarid rangelands

Charles Perrings and Brian W. Walker

6.1   Introduction

One of the main biological challenges to the economics of natural resource utilisation relates to the fact that almost all dynamic resource models are now being shown to ignore those properties of natural systems which most affect their evolution over time. The economic utilisation of natural resources involves the exploitation of ecological systems that turn out to be highly nonlinear, and characterised by discontinuous and sometimes irreversible change around a range of "thresholds." Yet most resource models either implicitly or explicitly assume the continuity, differentiability and reversibility of the underlying ecological functions. It has long been recognised that negative environmental externalities compromise the concavity–convexity conditions for social optima (cf Baumol and Bradford 1972; Starrett 1972), but the problem raised by the complexity of ecological systems goes well beyond the nonconvexity of the social possibility set. Discontinuities in the evolution of economically exploited ecological systems imply discontinuities in the economic system. That is, if economic pressure on any ecosystem causes it to "crash" – to undergo some fundamental and irreversible internal reorganisation – then the economic activities concerned will be similarly disrupted. Since many of the sources of discontinuous change in economically important ecological systems lie in the alteration of species diversity, this challenge is raised in a particularly acute way in the biodiversity problem.

This chapter considers the decision-making problem posed by the discontinuous change in of a group of ecosystems that underpin what is the basic economic activity of much of the population of Sub-Saharan Africa, and large parts of South Asia and Latin America. More particularly, it considers the implications of discontinuous change in the mix of species in semiarid rangelands for the optimal allocation of resources in the livestock sector. The notion that rangelands are characterised by discontinuous and

190

sometimes irreversible change is of comparatively recent origin. The theory of ecological succession that has dominated for much of this century (Clements 1916) in fact assumes all ecological systems to be globally stable around a unique climatically determined equilibrium state – the climax state. Moreover, convergence on the climax state has been assumed to be monotonic. The implication of this is that any perturbation of an ecological system will be followed by a return to the climax through a well defined succession of species. Indeed this is the basis of all existing range management models, and the rationale for such concepts as the carrying capacity of the range and the maximum sustainable grazing pressure. It is also the justification for stocking strategies based on the notion of a balance between successional tendency and grazing pressure.[1]

What range ecologists have begun to show is that this model of ecosystem dynamics has neither predictive nor prescriptive power. In fact, the semiarid rangelands are characterised by a wide range of management systems, from nomadic pastoralism to commercial ranching, and few of these reflect the equilibrium strategy authorised by Clementsian theory. Many such rangelands have been substantially altered as a consequence of the management regime (Noy-Meir and Seligman 1979), and almost all are currently argued to be degraded to a greater or lesser degree. This has led to a reappraisal of the environmental effects of different range management strategies (Perrings 1992). For example, whereas environmental economists have tended to argue that stocking levels under traditional "opportunistic" strategies are responsible for much of the rangeland degradation observed in the semi-arid regions of Sub-Saharan Africa (cf Dixon et al. 1989; Pearce et al. 1990), range ecologists have recently claimed that opportunistic range management techniques are not only more efficient than techniques based on equilibrium range succession models (Behnke 1985; Scoones, 1990; Westoby et al. 1989) but are much less environmentally damaging in such circumstances than is commonly argued (Abel 1990; Biot, 1990).

This chapter addresses the problem of optimal resource allocation under the discontinuous change in semiarid rangelands through states, each of which is characterised by a different combination of species – a different pattern of biotic diversity. More particularly, it constructs a non-Clementsian approach to the management of natural resources in such circumstances that is, hopefully, both intuitive and accessible. The chapter is in six sections. The following section discusses the ecology of semiarid rangelands, and identifies the role of biodiversity in the discontinuous eco-

---

[1] For a critique of the Clementsian model see Walker (1988), Westoby et al. (1989), Friedel (1991), and Laycock (1991).

logical change that is our main concern here. Section 6.3 then elaborates the first component of a simple decision model based on the state-and-transition approach to discontinuous change in semiarid rangelands: the state. Section 6.4 considers the second component: the transition. In both cases the treatment is kept simple by abstracting from important aspects of the evolution of rangelands. Section 6.5 considers the implications of this treatment and offers two extensions of the decision model to address these aspects. Section 6.6 illustrates the problem through simulations based on a specific model of livestock production which includes an environmental damage function, and a final section offers some concluding remarks.

6.2   Biodiversity and the ecology of change in semiarid rangelands

At a descriptive level, there are three aspects of the evolution of semiarid rangelands which reflect its complexity: the presence of lagged and frequently irreversible feedback effects; the sensitivity of the system dynamics to "rare" and extreme events; the existence of multiple locally stable states and paths of transition between such states across system thresholds. These are not mutually exclusive aspects, but it is convenient to use them to describe the role of biodiversity evolutionary processes of interest.

There are a wide range of feedbacks in any ecological system. Here we are concerned with those feedbacks in semiarid rangelands that (a) affect the economic productivity of the system by changing the mix of species, and (b) are related to the choice variables in the economic problem. More particularly, we are concerned with the feedback effects of grazing pressure on the nature and distribution of both vegetation and herbivory. The main determinants of that vegetation which make rangelands economically interesting – for grazing and browsing – are the soil moisture and nutrients available to plants. However, the dynamics of rangeland vegetation will, in addition, reflect the effects both of exogenous "events" – climatic fluctuations, fire and so on – and the forcing effect of herbivores, including livestock.

All three things – the determinants of natural vegetation, exogenous natural shocks, and the forcing effects of grazing pressure – are strongly interactive. There is, for example, a strong correlation between soil nutrients, soil moisture and soil texture; soil texture determining the degree to which both soil moisture and soil nutrients are available to plants. Sandy soils have a low water holding capacity but a high proportion of soil moisture is available to plants. Heavy clays are the opposite. Plants adapted to sandy soils have a high ratio of underground to aboveground biomass. Again, plants adapted to clay soils are the opposite. One implication of this is that grazing or burning affects a smaller proportion of the biomass of plants on sandy than on clay soils. But equally, the sandier the soil the greater is the competitive advantage of deep-rooted woody plants over shallow-rooted grasses (Knoop and Walker 1985).

To illustrate the role of biodiversity in the dynamics of semiarid rangelands we consider just two combinations of species: the balance between types of grass available to herbivores, and the balance between woody plants and grasses. Both are of immediate and obvious economic interest, and so are useful for our purposes – although we do not suggest that they are the only relevant diversity issues. Associated with each of these two issues is the spatial distribution or patchiness of species of all types. In semiarid regions with an incomplete vegetation cover, and for any given topography, there are three critical surfaces: an erosion or runoff surface, a transport surface, and a deposition surface (cf Pickup 1991; Tongway and Ludwig 1989). The first of these is the lower edge of any productive zone, where plant cover is inadequate to protect soil susceptible to water erosion. The second is a zone across which water and soil move with no net change in the soil. The third is the upper edge of a vegetated area, in which soil is deposited and water infiltrates. This deposition zone extends to the limit of water infiltration. The location of these zones is sensitive to climatic fluctuation. In dry years, the lower margin of a deposition zone will be starved of water with the result that the grass cover dies, becoming a new erosion surface in the next high intensity rains. The three zones accordingly "migrate" episodically upslope in a process that affects the distribution of both water and nutrients. This may change the dynamics of runoff by, for example, rill and gully formation, and so increase the risk of net soil loss, but its main effect is to create a mosaic of high-production patches with nutritious fodder in a larger, background area of low production, poorer quality vegetation. In the absence of field data Walker has conjectured that the process has two main implications, one positive and one negative. On the positive side, the high-production patches may provide higher quality food longer into the dry season than would occur in their absence. On the negative side, patchiness is associated with a change in the composition of species in favour of woody plants, which reduces the economic productivity of the range.

Consider, first, the general character of the process by which the mix of species changes in such systems. The Clementsian theory of plant succession implies that change in the composition of species in an ecosystem is incremental, or marginal, and is brought about by "biotic reaction": the modification by plants and animals of the abiotic environment. In the absence of macroclimatic change, the accumulation of biomass in any given system is held to lead to microclimatic changes, the alteration of soil organic matter and so on, and this in turn induces the plant succession. It turns out that in most rangelands this theory proves to be a poor predictor of change. Although Clementsian processes are undoubtedly at work, almost all significant changes are abrupt and discontinuous, and frequently in response to events or sequences of events of low probability. In the absence of such extreme events, net primary production varies in response to variation in rainfall, but it turns out that species composition remains essentially the same. The main reason advanced for the episodic nature of change in species composition is that both the establishment and mortality of plants is generally a function of a very particular set of conditions, which may only occur in association with "rare" events. So, for example, the establishment of cohorts of the perennial *Astrebla* grasses on the Mitchell grass plains of Australia is associated with particular phases of the El Nino Southern Oscillation, when good rains in late Autumn follow a year in which poor summer rains result in reduced vigour and abundance of the main competitor, *Danthonia* (Austin and Williams 1988).

What makes the problem of interest here is that the sensitivity of the composition of species in semiarid rangelands to extreme events is a function of economic decisions. That is, the level of grazing pressure affects both the distribution and the physiological status of plants, and through this their response to the events or fluctuations that drive the system. Put another way, there are feedback effects to the level of grazing pressure that work through event-driven change in the composition of species.

At the most general level, grazing pressure affects the sensitivity of the system to the levers of change: the extreme events. In terms of climatic fluctuations, for example, grazing pressure directly influences the relations between topography, rainfall and erosion, by altering the patchiness of rangeland. This has the effect of exaggerating both erosion and deposition (Stafford–Smith and Pickup 1990), so altering the response of the system to extreme climatic events. More particularly, it polarises the effects of extreme events. In source areas, it requires ever larger events to trigger a change.

The problem at issue here is the indirect effects of grazing pressure on the resilience of rangelands. If the recovery of species composition of rangeland following an environmental shock is becoming more sensitive to environmental fluctuations, then that rangeland may be said to be losing resilience. In other words, the smaller the event needed to force a system to undergo a self-reorganisation, the lower the resilience of that system. The argument is that a rise in grazing pressure can and does have the effect of lowering the resilience of rangelands to external stress, and so increasing the propensity for irreversible systemic change.

More particularly, an increase in grazing pressure implies a reduction in the proportion of palatable grasses and an increase in both unpalatable grasses and woody plants. The outcome of competition between palatable and unpalatable grasses, in response to an increase or decrease in grazing pressure, can lead to a hysteresis effect in the dynamics of their relative proportions (Noy-Meir and Walker 1986). The effect is induced by asymmetric competition between mature plants with well established root systems, and newly establishing plants. The result is that a reduction in grazing pressure leads to an increase in the proportion of palatable plants that is often much less than that expected by range managers (leading to disillusionment with standard recommendations for "improved" range management). Similarly, in many savannah rangelands woody plants are controlled by a combination of periodic fire and the capacity of the grass layer to suppress woody seedlings (Knoop and Walker 1985). In this case an increase in grazing pressure both raises the probability that woody seedlings will successfully establish, and reduces the effectiveness of fire as a control. Once established, the growth of woody plants generates positive feedback effects. By reducing grass cover more water passes through to the subsoil, taking nutrients with it, and enhancing the competitive ability of deeper rooted plants. In addition, by reducing the availability of "fuel," fire ceases to provide the check on woody plant growth that it does in a natural savannah.

The net effect is that the rangeland is transformed from one state (characterised by one mix of species) to another state (characterised by another mix of species). The transition between states occurs at particular threshold values for the system parameters: species populations, soil condition, degree of patchiness and so on. Neither the biological nor, in most cases, the economic productivity of the system is the same in different states.

Indeed, in most of the cases considered by dryland range ecologists, the change in the mix of species due to the effects of increased grazing pressure has involved a reduction in the productivity of the rangeland. The effects are not uniform, however, and as an example of an exception to this trend the grazing-induced change from saltbush shrubland to a grassland in South Australia involves no change in sheep productivity (Graetz 1986).

6.3   The "state" in the "state-and-transition" model

The state-and-transition model proposed by Westoby et al. (1989) attempts to conceptualise the discontinuous evolution of rangelands in a way that makes the management options transparent. In this section we consider a model that respects the intent of the state-and-transition model as a vehicle for the analysis of resource allocation under discontinuous change in semiarid rangelands. The basis of the state-and-transition model is the notion that there exist a number of rangeland "states," each characterised by a different composition of species, and each associated with a different set of future states. So, for example, a rangeland with viable seeds of woody plants in the soil is in a different state to that same rangeland without such seeds, because their potential future states differ. Put another way, the states of the rangeland are defined in terms of their future evolutionary potential. The transition between one state and any of the states with which it is associated is argued to be predominantly a function of exogenous events. Hence, the model is summarised by a list of potential states and the transitions between them, together with the conditions that induce transition from one state to another.

The main insights of the model are that evolution of the mix of vegetation species in semiarid rangelands occurs via discontinuous and often irreversible shifts from one "persistent" locally stable equilibrium state to another. This is not always the case, and there are examples of Clementsian-type gradual change from one recognisable state to another. Nevertheless, the assumption that each state is persistent within a range of external forces is the best overall assumption that we can make. Later, we will need to examine the consequences for optimal policies (derived on the basis of this assumption) of gradual transition in the absence of any significant events. Management action on rangeland in a locally stable equilibrium state – such as a change in the stocking rate, the deletion of some existing species or introduction of a new species, a change in the availability of water through dam construction or the drilling of wells, or a change in the composition of the soil through the addition of fertilisers, herbicides or pesticides – is argued to create the precondition for the evolution of the system by altering the sensitivity of the rangeland to exogenous natural shocks. Put another way, management actions change the stability of the equilibrium state before exogenous shocks. Transition from one rangeland state to another is then triggered by such shocks. That is, transition occurs when some locally stable equilibrium state is perturbed beyond the bounds of its stability. The process of transition is then argued to be largely independent of management action. It is driven by exogenous shocks that are independent of herd density or grazing pressure. So, for example, Scoones (1990) has argued that in systems "away from equilibrium" density-independent factors such as rainfall dominate herd density in explaining herd dynamics.

The important points here are that persistent or equilibrium rangeland states are locally, not globally, stable. That is, if a system at some equilibrium is sufficiently perturbed, it will not reconverge on that equilibrium, but will instead converge on some other equilibrium. It will be irreversibly transformed from one persistent state to another. Moreover, the limits of the local stability of any given equilibrium (in the absence of further management intervention) will define threshold values for livestock, vegetation, water flows and so on. If any one of these variables is driven beyond such threshold values, the system itself will undergo irreversible change. At a conceptual level it is helpful to separate the persistent states or locally stable equilibrium systems from transient states or unstable systems. Since the system parameters and dynamics differ both from one stable equilibrium to another and since the role of endogenous and exogenous forces in system dynamics differs between stable and unstable equilibria, they involve qualitative differences in the resource allocation problem. Accordingly, we distinguish two classes of decision problem: those associated with locally stable equilibrium rangeland states, and those associated with the transition between states.

With respect to the first class of decision problems, the optimal allocation of resources is that which maximises an appropriate measure of welfare through choice of stocking density, subject to the dynamics of both the natural and economic environments within which the decision takes place. The solution to the problem may be said to be ecologically sustainable if it does not drive the system beyond the bounds of local stability. A stocking strategy may be said to be ecologically unsustainable if it does – implying that the system loses resilience in the face of climatic shocks as a consequence of the stress to which it is subject at the optimal level of economic activity. This is an interpretation that is similar to that of Conway and Barbier (Barbier, 1989c; Conway 1987; Conway and Barbier 1990), and directs our attention to the dynamic relation between the economic environment which determines optimal grazing pressure, and a stochastic natural environment which is the source of shocks or natural events (Perrings 1993).

To simplify the problem as much as possible, the mix of species on the range is summarised by a single measure of carrying capacity. This implies that the diversity of species is of interest only to the extent that it is instrumental to the objectives of livestock farmers. Clearly this is much too restrictive given the public good nature of diversity, but it serves our interest in the effect of a particular set of feedbacks between biodiversity and stocking densities. More particularly, it serves our interest in the feedback effects of current stocking densities on the future carrying capacity of the range, where carrying capacity also depends on variance in rainfall. It follows that the available vegetation is not independent of the history of rangeland use, and current herd size is not independent of the past evolution of available vegetation. The underlying growth functions for the herd and the vegetative cover of the range are assumed to be logistic. The relation between carrying capacity and grazing pressure is defined within the model. The relation between carrying capacity and variance in rainfall is reflected in the variation of the ecological parameters of the model. Herd growth, the impact of herd size on rangeland vegetation, and the recuperative powers of the range are jointly determined, and are all sensitive to fluctuations in rainfall.

Formally, we assume that there exist n possible rangeland states, S_i, $i = 1, \ldots, n$. To each of these states, there corresponds a time path for both herd size and vegetation that depends on the initial values of each at the inception of the state. Vegetation is defined, as we have already remarked, in terms of carrying capacity and so is measurable in livestock units. Assuming discrete time, the time paths for the size of the herd, $x(t)$, and the carrying capacity of the range, $k(t)$, in state i are described by the sequences $\{x_i(t_i)\}$ and $\{k_i(t_i)\}$. It is convenient, to start with, to consider only the problem of resource allocation in the prevailing state, ignoring the possibility that it may transmute into some other state at some point in the future. It is convenient, that is, to start with something that looks very much like a traditional range succession problem. For the moment, therefore, we suppress the state subscript, i.

Although this rules out the discontinuous change that marks the evolution of a system through different states, we still have to deal with feedback effects. The time paths for both herd size and carrying capacity in any state are affected by the history of stocking and offtake decisions in that state. That is, if "growth" in both the size of the herd and the carrying capacity of the range in the prevailing state is summarised by some set of forward recursive functions (the equations of motion of the system), the arguments of those functions will include all the current values of all the variables that affect the productivity of the system. In this simple illustration we assume that the forward recursions take the general form:

$$x(t + 1) - x(t) = f[x(t), k(t), u(t)] \tag{6-1}$$

$$k(t + 1) - k(t) = g[x(t), k(t), u(t)] \tag{6-2}$$

$x(t)$ and $k(t)$ have already been defined; $u(t)$ denotes offtake at time t; and the functions $f[\cdot]$ and $g[\cdot]$ describe the net growth rate of the herd on the range and the net rate of regeneration of the range respectively. Although we specify only general forms of these functions, in any specific form the net growth of the herd would be equal to the difference between the natural growth of the herd during that period – given the degree of grazing pressure, $x(t)/k(t)$, and the state of the range – and the level of offtake. Similarly, the net growth in the carrying capacity of the range would be equal to the difference between the natural regeneration of the range – again, given the degree of grazing pressure, $x(t)/k(t)$, and the state of the range – and the depletion of carrying capacity through grazing/browsing of palatable grasses, shrubs and trees. The relationship between herd growth, carrying capacity and offtake in any particular rangeland state would reflect the feedbacks discussed earlier, but in general offtake would have both direct and indirect effects on the size of the herd. If livestock were drawn off in the current period, the current size of the herd would be reduced. At the same time, however, the future growth potential of the herd would be improved due to the positive feedback effect of lower herd sizes on the carrying capacity of the range.

It is assumed that all variables are subject to random fluctuation depending on the variation in rainfall. In general, the rate of herd growth and the rate of range regeneration would be expected to vary directly with rainfall, and the rate of range depletion due to grazing would be expected to vary inversely with rainfall. However, as has already been remarked, empirical research has shown that the time behaviour of such recursions is often extremely complex. The herd growth function, for example, exhibits normal compensatory, overcompensatory, depensatory and critical depensatory properties for similar herd sizes at different periods. Range depletion resulting from low rainfall (through mortality of tillers and whole plants) occurs very quickly, in one season, but regeneration in response to a return to the previous level of rainfall usually takes longer (two or more seasons). This is one aspect of the hysteresis effect and is the reason why the long-term sustainable stocking of rangelands is less than that predicted by average rainfall. The higher the coefficient of variation in rainfall, the greater is the discrepancy. Nevertheless, the assumption that we are dealing with persistent or locally stable equilibrium states does imply that in the absence of grazing pressure and extreme natural "events," the carrying capacity of the range will converge to some positive equilibrium value. In other words, in any locally stable equilibrium state, there exist a set of ecological parameter values at which the state will "persist."

The system is "controlled" through the level of offtake, $u(t)$, which is assumed to be a function of the economic environment within which decisions are made. That is,

$$u(t) = h[p(t)] \qquad (6\text{-}3)$$

in which $p(t)$ summarises the relevant terms of trade at time t. For the moment we make the standard assumption that the resource user is unable to influence either the prices of inputs or the price of output. Hence $p(t)$ is independent of the histories of $x(t)$, $k(t)$ and $u(t)$ and so is insensitive to change in the natural environment as a result of private resource use. That is, the degradation of the rangeland will not feed back into relative prices. This assumption will jar most readers. We will, however, be returning to it later. What it means for now is that the economic problem for the private resource user involves the maximisation of some index of welfare over some defined time horizon through choice of the level of offtake, and subject to the properties of a physical system that is sensitive to the level of resource use, and an economic system that is not. That is, assuming a time horizon of T periods, the problem is to:

$$\max_{u(t)} E\sum_{0}^{T} r^t w[x(t), k(t), u(t)]$$ (6-4)

subject to

$$x(t + 1) - x(t) = f[x(t), k(t), u(t)]$$ (6-5)

$$k(t + 1) - k(t) = g[x(t), k(t), u(t)]$$ (6-6)

$$u(t) = h[p(t)]$$ (6-7)

$$x(0) > 0 = x_0$$ (6-8)

$$k(0) > 0 = k_0$$ (6-9)

$$x(t), k(t) \geq 0$$ (6-10)

Expected welfare, $Ew[\cdot]$, is a function of both the state of the assets, $x(t)$ and $k(t)$, and the consumption of the herd, summarised by $u(t)$. It is indirectly a function of the economic environment, $p(t)$. $r = [1/(1 + d)]$ is a discount factor with d being the rate of discount. Since we are describing only the prevailing state, the initial values $x(0)$ and $k(0)$ are assumed to be given.

The solution to this problem provides a decision rule which fixes the optimal offtake policy under the particular rangeland state. It requires, as one would expect, that the marginal benefit of any stocking level (herd size) should equal the marginal cost of that stocking level, given the set of relative prices and the rate of discount that comprise the economic environment.[2] The optimal offtake is then that level of offtake which adjusts the herd size to its optimal value – assures the optimal level of grazing pressure – for the current state of the rangeland. If the optimal level of

---

[2] The first order conditions for an optimum require that:

$$0 = \quad H_{ut} = w_u(t) - \rho\lambda(t+1) f_u(t) + \rho\zeta(t + 1)_{u(t)}$$
$$\rho\lambda(t + 1) - \lambda(t) = \quad - H_{x(t)} = - w_x(t) - \rho\lambda(t + 1) f_x(t) - \rho\zeta(t + 1)g_{x(t)}$$
$$\rho\zeta(t + 1) - z(t) = \quad - H_{k(t)} = - w_{k(t)} - \rho\lambda(t + 1)f_{k(t)} - \rho\zeta(t + 1)g_{k(t)}$$
$$x(t + 1) - x(t) = H\rho\lambda_{(t+1)} = \quad f[x(t), k(t), u(t)]$$
$$k(t + 1) - k(t) = H\rho\zeta_{(t+1)} = \quad g[x(t), k(t), u(t)]$$

$$x(0) = x_0$$

$$k(0) = k_0$$

$$u(t) \in U$$

where

$$H[x(t),k(t),u(t),\lambda(t)] = w[x(t), k(t), u(t)] + \rho\lambda(t + 1)f[x(t), k(t), u(t)]$$
$$+ \rho\zeta(t + 1)g[x(t), k(t), u(t)]$$

is the current value Hamiltonian for the problem, and λ and ζ are Lagrangian multipliers.

grazing pressure is such that the corresponding state of the rangeland is resilient in the face of natural shocks for all t, this corresponds to the solution that would fall out of a range succession or Clementsian model. But there is nothing in the problem to assure that this will be so.

There are two senses in which it is meaningful to describe a range as overgrazed under some strategy. Overgrazing in an economic sense may be said to exist wherever the actual level of grazing pressure exceeds the optimal level of grazing pressure. Overgrazing in an ecological sense may be said to exist wherever the actual level of grazing pressure exceeds the level at which the system is resilient. Whether optimal grazing pressure is sustainable under some particular state depends on the economic parameters of the system. If relative prices are such that it is optimal to "mine" the range, then it will be ecologically overgrazed at the economically optimal level of grazing pressure. In terms of the biotic diversity of the range, what this means is that the higher the level of grazing pressure the more sensitive will the biota of the rangeland become to variation in rainfall or temperature. In terms of the stability (as opposed to the resilience) of the range, the higher the level of optimal grazing pressure relative to the rate of range regeneration under some persistent state, the less stable will be that state, and the greater the probability that normal climatic fluctuation will trigger a change of state.

6.4   "Transition" in the "state-and-transition" model

To address the decision-making problem of the discontinuous evolution of rangeland takes us beyond the existence of a given state, and requires that we include both future possible states other than the prevailing state, together with the "transient states" between future possible states – the unstable equilibria that mark the transitional phase of a system in the process of transmuting from one locally stable equilibrium state to another. The argument of range ecologists is that at unstable equilibria convential range management is a blunt instrument. Density-independent factors have a much greater role to play in the evolution of the system than at other times. This is not to say that stocking decisions are irrelevant to the outcome at unstable equilibria. There is an interaction between management and density-independent environmental effects which will influence subsequent dynamics. For example, where a bad drought leads to high animal mortality, subsequent regeneration of the perennial plant population benefits from a low stocking rate. The animal populations increase at a slower rate than the plant populations. Where animal offtake early in the drought reduces peak-drought grazing pressures, such that animal population crashes are preempted, the establishing plant populations post-drought are subject to higher grazing pressure. This both slows regeneration and, through selective grazing effects, influences the composition of plant species. On the other hand, establishment success of woody plants within the grass sward after a drought will depend heavily on the level of active grass cover. Death of perennial grasses during the drought significantly increases where grazing pressures are high (see Walker et al. 1986). However, the outcome will be particularly sensitive to exogenous shocks at this time. It follows that if stocking decisions at the prevailing state cause the system to lose stability, then one cost of the decisions is the loss of control that occurs in the neighbourhood of unstable equilibria. It also follows that if stocking decisions threaten the stability of the prevailing state, the cost benefit calculus on which any current decision is based should include estimates of the costs and benefits under all states that are reachable from the current state.

In one sense the problem looks very familiar. We are, after all, used to thinking of decision-making under uncertainty in terms of the maximisation of some measure of well-being as a function of state-contingent bundles of commodities. Take a simple case. Suppose that there are only two possible stocking strategies corresponding to each state of the rangeland, and that each strategy involves a different optimal level of grazing pressure. If the transition possibilities for each state are a function of the stocking strategy applied plus some random event or set of events, and if under any given stocking strategy the rangeland state may either persist or transmute into some other state, then the model can be represented by a simple decision tree such as that shown in Figure 6.1.

In this figure the rangeland states, S_i, form decision nodes, to which correspond stocking strategies P_{ij}. Given an estimate for the probability of the events or shocks that drive the system, and given the sensitivity of any state to such events under each stocking strategy, the probability that a state will persist or transmute under a given stocking strategy will be known. The expected value of all stocking strategies and hence the optimal policy will be identifiable.

Nor is this a wholly misleading way of thinking about the problem. If the choice of stocking strategy alters the probability of a change in the state of the range, i.e., if there are environmental feedbacks in the system, then the rangeland state itself becomes a choice variable, if only in a probabilistic sense. The decision tree makes this transparent. It also makes transparent the irreversibility of transition between states, and so the conditions for the reversibility of stocking strategies. The chance nodes (P) in this formulation of the problem approximate the transient states. The implication is that once a stocking strategy has been selected, the probability that it will flip the system from the prevailing state to some other state

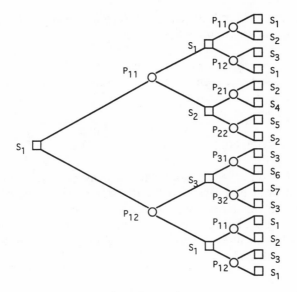

Figure 6.1. State-and-transition decision tree.

is knowable providing (a) that the set of possible states and the rangeland dynamics in each state are known, and (b) that the distribution of both economic and natural environmental conditions is known.

In terms of our earlier discussion, the decision problem under discontinuous change of this sort may be written in the following form:

$$\max_{u(t)} E \sum_{i=1}^{n} \sum_{t_i=0}^{T_i} r^{t_i} w[x_i(t_i), k_i(t_i), u_i(t_i)] \tag{6-11}$$

subject to

$$x_i(t+1) - x_i(t) = f_i[x_i(t_i), k_i(t_i), u_i(t_i)] \tag{6-12}$$

$$k_i(t+1) - k_i(t) = g_i[x_i(t_i), k_i(t_i), u_i(t_i)] \tag{6-13}$$

$$u_i(t) = u_i[p(t)] \tag{6-14}$$

$$x_i(0_j) = F_i(S_j) \qquad j = 1, \ldots, i-1 \tag{6-15}$$

$$k_i(0_j) = G_i(S_j) \qquad j = 1, \ldots, i-1 \tag{6-16}$$

$$x_i(t_i), k_i(t_i) \geq 0 \tag{6-17}$$

That is, welfare is maximised by choice of stocking rate over a time horizon equal to the sum of the expected "lives" of all states considered in the decision problem, and subject to (i) the rangeland dynamics in each state of Equations (6-12) and (6-13); (ii) the dependence of offtake on the economic environment Equation (6-14) (assumed here to be independent of all states); and (iii) to transition functions (6-15) and (6-16) which specify the initial values of $x_i(t_i)$ and $k_i(t_i)$ as a function of the preceding states.

The necessary conditions for the optimisation of the problem with respect to each state are identical to those already discussed. They include equality between the marginal costs and benefits of livestock holdings under the "optimal" state. This admits the possibility that the system will pass through a number of states before attaining the optimal state (or states). Since $x_i(0_i)$, $k_i(0_i) = 0$ if the ith state is not generated by the preceding states, the Functions (6-16) and (6-17) serve to trigger the inclusion of that state. It is only if $x_i(0_i)$, $k_i(0_i) > 0$ with some positive probability that the ith state will be relevant to the decision-making process. We note (without comment at this point) that since the terms of trade, $p(t)$, are not sensitive to the evolution of the rangeland this implies that optimisation of the problem requires that the state of the range be adjusted to the set of relative prices and not vice versa.

6.6   Assessment and extensions

Semiarid rangelands offer insights into three important aspects of the biodiversity problem. The first concerns the identification of a problem that is much wider than the extinction of species in tropical moist forests. Without in any way diminishing the potential costs of species extinction, it is worth underlining the fact that the greatest costs of a change in biodiversity to this and the next few generations are likely to lie in changes to the mix of species in a wide range of tropical and nontropical ecosystems, relatively few of which involve the global extinction of species, but all of which reduce the value of the ecological services derived from the natural environment. A shift in the balance between palatable and unpalatable grasses in the rangelands of Sub-Saharan Africa is as much a part of the biodiversity problem, and may be more a part of the development problem than, say, the global extinction of a species of insect with a very narrow niche in the tropical moist forests of Central America. The first set of insights relate to the long-term significance of biodiversity change that involves much less than species extinction. All this does not lessen the significance of global extinctions, both in their own right and as evidence of serious environmental and management problems. What it does is to bring to the fore another aspect of biodiversity loss that deserves equal attention.

The second set of insights concerns the interaction between the ecological and economic parameters of the system. It turns out that whether fluctuations of a given magnitude in the set of relative prices are destabilising depends on the variation in the ecological parameters of the system, and vice versa. This set of insights relates to the implications of discontinuous change for the price system. The third set of insights concerns the problem of fundamental uncertainty caused by irreversible but essentially unforeseeable change.

The examples of irreversible change in both the biological and economic productivity of semiarid rangelands offered in this chapter for the most part involve something less than the deletion of any species. The switch from palatable to unpalatable grasses, or substitution of woody plants for grasses both involve an irreversible change in the state of the range that have major implications for its productivity and so for the welfare of current and future users of those rangelands.[3] The driving forces behind such changes are very similar to those behind the more dramatic examples of biodiversity loss elsewhere. Exploitation of the resource base in the form of livestock farming places the supporting ecological systems under increasing levels of stress, which in turn increases their sensitivity to natural shocks or "events." This implies the weakening of the resilience of those ecosystems, and opens up the prospect of their transmutation. In many cases an associated cause of both the change in the mix of biodiversity and the loss of system resilience is soil degradation. Soil degradation is for the most part irreversible, and also for the most part implies a decrease in the biological productivity of the system. Hence if grazing pressure destabilises successive rangeland states through the degradation of the soil, it also ratchets downwards the productive potential of the range – leaving all future generations worse off. It follows that the degradation of semiarid rangelands has the potential to worsen the intergenerational distribution of assets and income in just the same way as does the global extinction of species.

The second set of insights bears on the assumption made that individual resource users can have no impact on relative prices through their actions. In this paper we have ignored effects of grazing that are visited on users of rangeland resources other than livestock farmers, but we have made the standard competitive economy assumption that individual resource users have no effect on the terms of trade. Yet the conclusions coming out of the state-and-transition model suggest that individual actions may have an enormous impact on the productive potential of the system wherever they prove to be destabilising. The "straw that breaks the camel's back" has a direct analogue in the stocking decisions of pastoralists. The stocking decisions of individual resource users may indeed

[3] Note that "irreversible" means ecologically irreversible in the relevant time frame. The change may well be reversed through mechanical or chemical intervention, and given enough time slow evolutionary processes can also lead to reversal.

affect the productivity and hence the value of the range. But if this is not reflected in the set of relative prices on which users base their decisions (as it cannot be in pastoral systems based on traditional communal property rights, for example), the result will be an intertemporally inefficient allocation of resources. The private decisions of pastoralists will be based on a calculus of costs and benefits that ignores the user component of the social cost of the resource.

There will always exist a set of relative prices and/or discount rates at which it will be privately optimal to destabilise the rangeland whatever the state it is currently in. Moreover, just as the ecological parameters of the system were argued to be interactive, so too are the economic parameters. A price change that causes a strategy to "crash" under one discount rate, for example, will not necessarily do so under another. This relationship between the set of relative prices, the rate of discount, and the degradation of environmental resources has been exhaustively discussed in the theoretical literature on natural resource utilisation and need not detain us here. What is worth spending a moment on is the basis for the assumption that the terms of trade are insensitive to changes in the value of the resource base, and the implications of this for the decision process. The main propositions coming out of the literature on the link between economic incentives and the degradation of semiarid rangelands are the following:

1. that the depression of prices for marketed agricultural output by governments has historically increased the degradation of soil resources by reducing the value of those resources and so the incentive to invest in conservation;
2. that the subsidisation of (marketed) agricultural inputs has had the effect of encouraging the overutilisation of those inputs, and
3. that the lack of secure and well defined property rights in rangelands has created at least one of the conditions for open-access common property problems of the "tragedy of the commons" type.[4]

Taking 1 and 2, it is intuitive that an increase in the net producer price of livestock ($w_{u(t)}$ in the general form of the benefit function used here) will reduce grazing pressure by increasing the net benefits of offtake. If the cost of livestock holdings and of access to grazing land are held constant while the producer price is increased, the short run effect will be to stimulate offtake. Hence the depression of producer prices is argued to have weakened the incentive to destock the range. This is especially significant in the low income countries where the net benefits of livestock

---

[4] See, for example, Pearce et al. (1988); Repetto (1986, 1989); Perrings (1991); Cleaver (1988); Warford (1989); Barbier (1988, 1989b); Markandya (1991).

holdings (inventory) tend to be much higher than elsewhere. Specifically, since livestock offer major benefits in such countries in terms of animal products and services, such as draft power, collateral in credit markets, and so on, the optimal level of grazing pressure will tend to be higher than elsewhere for a given set of producer prices. Depression of producer prices in those countries will exaggerate this difference. So too will depression of the cost of livestock holdings, and despite the trend towards the liberalisation of agricultural markets in recent years, the evidence for distortions of both kinds continues to be very strong. While producer prices have tended to converge on international prices, subsidies on water provision, fodder, veterinary services, fencing and so on remain in almost all semiarid countries, and have been argued to contribute to leaching, soil acidification, and loss of soil nutrients, and to the reduction in the resilience of key ecosystems (Grainger 1990).

Taking 3, it is a fact that most rangeland in the low income countries continues to be used on the basis of some form of communal tenure. There are accordingly no markets to signal the change in value of either the land or the usufructual rights enjoyed by users. At the same time, the institutions traditionally charged with the allocation of usufructual rights and with the regulation of the level of activity have either disappeared or been severely weakened, while replacement institutions have failed to exercise the authority vested in them. Although there is still an incentive to cooperate where land is held communally (cf Cousins 1987; Wade 1987), wherever traditional institutions for regulating the use of communal lands have broken down individual pastoralists have tended to overgraze the range (Jamal 1983). In addition, the security of tenure over usufructs has diminished with the weakening of institutions of collective control (Bruce 1988). The net result is that there are neither implicit nor explicit signals of the social opportunity cost of land use. There exist neither prices nor regulatory mechanisms that are sensitive to the state of the range.

Since the feedbacks that give rise to the discontinuous evolution of the range indicate that there are long run costs to stocking decisions in terms of foregone productivity, the price set which underlies the term of trade variable $p(t)$ should be sensitive to the state of the system. Put another way, intertemporal efficiency in the face of these environmental feedbacks requires that $p(t)$ be related to $x(t)$ and $k(t)$ for each state of the range. At minimum, it requires that

$$p_i(t) = h_i[x_i(t_i), k_i(t_i)]$$ \hfill (6-18)

such that the marginal cost of a change in herd size or carrying capacity includes the indirect effects of the change in offtake stimulated by the price variation to which it gives rise.[5] At present the relative prices governing the private decisions of resource users in semiarid rangelands do not capture these feedback effects, and this remains a major source of difficulty for both management and policy.

The third set of insights deriving from our treatment of semiarid rangelands also relates to the pricing problem. It concerns the treatment of fundamental uncertainty. While the formulation of the decision problem offered in Section 6.5 provides a convenient way of thinking about the optimal allocation of resources over a sequence of states, it does assume that the set of possible future states of the range and the dynamics of each future state are knowable in advance – at least in a probabilistic sense. If there are no surprises in the future evolution of the range, then discontinuous change poses no fundamental difficulty for decision-making. While part of the motivation behind the state-and-transition model is the belief that enough is known about the transition between locally stable rangeland states that it is possible to improve the decision-making process very substantially, it would be misleading to imply that the discontinuous evolutionary change of any ecosystem is completely knowable in this sense. The overturning of the traditional range succession model reflects improvements in the understanding of rangeland dynamics. The data set on rangeland dynamics is much richer than it was even a decade ago. However, the more that is learned about the complexity of the system, the less that can be said in detail about the evolutionary path of that system following some destabilising perturbation. As described earlier, a change in state is very often the result of a particular sequence of events, in which exogenous environmental shocks have unusual significance because of the instability of the transitional state. The consequences of a major fire, for example, are highly sensitive to the post-fire rainfall regime. The outcome of stocking decisions made at that point is accordingly more fundamentally uncertain than at other times.

In any ecosystem having evolved a particular self-organisation, there exist values of the resources involved which are "thermodynamically" possible but which are incompatible with that self-organisation. Whenever the

---

[5] If prices were sensitive to the size of the herd and the state of the range, the first order conditions (suppressing state subscripts) would include the adjoint equations:

$$\rho\lambda(t + 1) - \lambda(t) = - w_{x(t)} - \rho\lambda(t + 1)[f_{x(t)} + f_{u(t)}u_{p(t)}h_{x(t)}] + \rho\zeta(t + 1)[g_{x(t)} + g_{u(t)}u_{p(t)}h_{x(t)}]$$

$$\rho\zeta(t + 1) - \zeta(t) = - w_{k(t)} - \rho\lambda(t + 1)[f_{k(t)} + f_{u(t)}u_{p(t)}h_{k(t)}] + \rho\zeta(t + 1)[g_{k(t)} + g_{u(t)}u_{p(t)}h_{k(t)}]$$

the second terms in each of square brackets on the RHS indicating the indirect effects of a change in the terms of trade on the evolution of the shadow price of the herd and carrying capacity of the range.

resources of such an ecosystem are driven past certain threshold values, the system will switch from one self-organisation to another. Threshold values exist, for example, for the diversity of species in an ecosystem. If any one population in an ecosystem falls below its critical threshold level the self-organisation of the ecosystem as a whole may be radically and irreversibly altered (cf Pielou 1975). Given the limitations on available data, and on our understanding of system functions, changes to the self organisation of ecosystems that lose resilience or local stability are not *a priori* predictable. A historical record of the evolution of semiarid rangelands in different circumstances has enabled the construction of a list of possible rangeland states and sequences of states, but the list is not exhaustive. In other words there remains a considerable amount of fundamental uncertainty about the long run implications of changes in rangeland states for which there are no precedents. The set of all possible states, and the dynamics of those states is not knowable – even probabilistically.

The problem of decision-making in these circumstances has been widely considered in the context of environmental questions generally and the biodiversity problem in particular. It has led some to argue that where the potential costs to future generations of the loss of system stability is conjectured to be unacceptable, and where the price system is such that private users have no handle on such potential costs, intertemporal decisions should be bound by a requirement for the preservation of at least components of the natural capital stock (cf Barbier, Markandya and Pearce, 1990; Pearce 1987, 1988; Pearce, Markandya and Barbier 1989; Pearce and Turner 1991) – at least until society is judged to have a clearer understanding of what the optimal level of resource stocks may be. What such a restriction implies in terms of the decision model described in Section 6.5 is a constraint on stocking densities so as to maintain populations/resource stocks within bounds that are judged to be consistent with the stability of the ecosystem concerned. For the ith state of the range this would take the form of an inequality constraint of the general form

$$r_i[x_i(t_i), k_i(t_i), u_i(t_i)] \leq 0 \quad i = 1, \ldots, n \quad 0 \leq t_i \leq T_i \qquad (6\text{-}19)$$

based on physical indicators of changes in overall regressive succession, standing crop biomass, relative energy flows to grazing and decomposer food chains, mineral nutrient stocks and the mechanisms of and capacity for damping oscillations (di Castri, 1987; Schaeffer et al. 1988).

The imposition of such a constraint means that the first order conditions for an optimum of the problem have two forms, depending upon whether the constraint is or is not binding. If the constraint is not binding the conditions are the same as before. If the constraint is binding, implying that the system is close to the margin of resilience in respect of some physical indicators, then optimisation of the problem requires that the cost of breaching the restriction be taken into account. In policy terms, if the constraint is nonbinding it implies free access to the resource in question. If the constraint is binding it implies restricted access to the resource secured by formal physical limits (quota, permits, safe minimum standards) or, more precisely, by the penalties associated with the violation of those limits.

6.7   Concluding remarks

This chapter has considered the decision-problem raised by discontinuous biotic change in semiarid rangelands. Discontinuous change is conceptualised as the transition from one locally stable rangeland state (to which corresponds one equilibrium mix of species) to another. The transition from a given state implies the loss of stability or resilience of that state. The proximate cause of transition is an exogenous environmental shock that perturbs the system beyond the limits of local stability, but the limits are themselves a function of range management decisions. The policy implications of the approach follow from this. Where the range is close to the limits of local stability, or where range management decisions have the effect of bringing the range close to the limits of local stability, there are grounds for introducing incentives or regulations to protect those limits.

PART III
ECONOMIC ISSUES

CHAPTER 7

Economic growth and the environment

Karl-Göran Mäler[1]

7.1　Growth and environment

Since the end of the 1960s there has been an ongoing discussion whether continued economic growth is compatible with finite supply of nonrenewable resources and with the sensitivity of renewable resources to the human interferences in natural ecological systems. On the one hand, environmentalists (and others such as the Club of Rome) have argued that the finiteness of the resource base must put an end, not only to continued growth but to the present lifestyles in the industrialised world. On the other hand, economists (but not all) have argued first that there are sufficient substitution possibilities between man-made capital and natural capital to make further growth possible and second that in addition to this, technical progress will reduce the dependence on natural resources. To this the environmentalists respond that the substitution possibilities are limited and technical progress cannot change the fundamental laws of thermodynamics. These are some of the issues to be discussed in this paper.

Irrespective of the prospects of future technological development or the substitution possibilities, we want to manage the use of environmental resources as well as possible. It is more important to discuss what is growing than whether we should have growth or not. Usually, growth is discussed in terms of GNP (or GNP per capita). But this we know is not an appropriate measure of welfare. How could we define a better index of growth? This is discussed in Section 7.4. But then, the question arises whether one could maximise the growth of such an index. This is discussed in a simplified overlapping generation model in the final section, Section 7.5.

[1] I am grateful to Andrea Beltratti for valuable comments. A different version of this chapter appears in Passinetti and Solow (1994).

213

7.2   Materials balance[2]

In elementary textbooks, a picture illustrating the circulating flow of goods and services is usually included in order to introduce the student to the interdependence of different markets in the economy. However, this figure usually neglects the circular flow between the human society and nature. Figure 7.1 illustrates the flow between the "conventional" economy and the supporting ecosystems.

The arrow called "inputs" illustrates the use of inputs from the environment in production. These inputs include mineral ore, oil, timber, fish, agriculture products, air, water etc. They are processed and turned into consumer goods and sold to households or turned into capital goods and invested in man-made real capital. The box called "production" illustrates this.

In this connection, it is important to remember the first law of thermodynamics, which states that matter and energy cannot be destroyed but only transformed into other kinds of matter or energy. This means that the *weight of the inputs entering the production box must equal the weight of the outgoing flows.* Corresponding to the mass entering production, there must be a flow with exactly the same mass leaving production. One component in this flow is the output of final goods – consumption and production goods. The other component is the flow of residuals or wastes which is "dumped" in the environment.

The same argument can be applied to the box called households. The flow of consumer goods entering this box must have a corresponding flow of residuals leaving the box (neglecting the mass stored in consumer capital). Some of these residuals are transported to waste treatment plants or to recycling plants. However, the important thing to remember is that waste treatment does not reduce the mass of the stream but only transforms it into other forms of matter which hopefully are less damaging to the environment. It is only in the capital box that the weight of the inflow of mass exceeds the outflow because of the storage of matter in real capital. Real capital was considered to be stored labour by the "Austrians." It can as well be considered as stored natural resources. The accumulation of capital can thus, at least for a while, break the connection between the ingoing flow of mass and the outgoing flow.

Besides providing inputs to production processes, nature or the environment also provides a broad variety of services directly to households. These services includes various life-supporting services – clean air, water,

[2] The material balance principle was first introduced into economics by Allen Kneese and Robert Ayres (Ayres and Kneese 1969) and was analysed by d'Arge, Ayres and Kneese (1970) and Mäler (1974). The following discussion is mainly based on Mäler (1974).

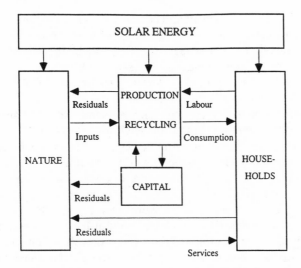

Figure 7.1. Flows between the economy and supporting ecosystems.

climate control, food etc. but also other kind of services like various amenities. In general, but not always, will the discharge of residuals affect the quantity and quality of these services? It follows that these services will depend on the amount of matter we extract from nature.

The whole system is driven mainly by current and past solar radiation (the exception being the energy released from nuclear plants). The concentrated solar energy is used by natural systems and by humans and is degraded by this process (the second law of thermodynamics). We will come back to this shortly.

There are several lessons to be drawn from this exercise with the material balance principle. First, an increase in the scale of the production will, unless there have been structural changes, imply an increase in the mass of inputs and therefore also an increase in the load on nature from discharging an increased amount of residuals. Only if economic growth takes the form of less "material intensive" technology will the damage to the natural environment be controlled. It seems that further recycling may also reduce the load on the environment. However, recycling will require energy and if that energy is coming from fossil fuels, then we are back to the original problem. We will increase the load on the environment.

The only sources of energy that may solve this problem are nuclear energy or solar energy. Nuclear energy has its own problems which will not be discussed here and we are left with solar energy. If technological progress may result in efficient and effective ways of collecting solar energy (and ways of exporting the high entropy heat resulting from the use of that solar energy to outer space), recycling may be the way of expanding the production without damaging the environment.

Thus, uniform growth implies worsened environmental conditions. But growth also implies a higher material standard of living which will, through the demand for a better environment, induce changes in the structure of the economy, so that the environment is less damaged. Thus, one would expect that a country in the beginning of its economic development will be experiencing a worsening of the environment, while a country in which growth has taken place over a longer period of time will be adjusting its patterns of growth in such a way that the environment in fact improves. A casual reading of the World Development Report 1992 indicates that this hypothesis is consistent with at least cross sectional data. Furthermore, in Panayotou (1992), the hypothesis is statistically tested (against cross-sectional data) and it is found that for quite a wide spectrum of environmental problems it is true that these become more serious with growth in the first phase of development but are very much reduced in later phases of development.

7.3   How should growth be measured?

Traditionally, growth is being measured as the relative change in GNP per capita. The basic idea is that GNP per capita measures the current and the future per capita welfare. GNP equals (in a closed economy) consumption plus gross investment. Current welfare is then measured by current consumption and the change in future consumption opportunities caused by current savings – and investment decisions are measured by current investments. Obviously, GNP is not an appropriate measure of welfare. Besides the fact that it does not incorporate equity issues, it also suffers from other drawbacks. In this connection, the following seem to be the most important failures:

1. GNP is not a net concept. We are interested in net investments and therefore in NNP.
2. The conventional net investments do not include changes in most natural assets, although, they are indispensable for human welfare.
3. The consumption as conventionally measured does not include environmental damage.

These, and other perceived drawbacks with GNP as a welfare measure have created a number of studies on how to redefine the Standard National Accounts in order to improve the informational content of NNP. United Nations Statistical Office has also produced a draft *Handbook on National Satellite Accounting*. However, most of these studies and recommendations are based on rather ad hoc assumptions and are frequently inconsistent with basic economic theory.

Dasgupta and Mäler (1991), Hartwick (1990), Hartwick (1992) and Mäler (1991b) have developed a theoretical framework, on which an accounting system may be satisfactorily based. It turns out that the "correct" measure should be constructed as follows:

NNP = Consumption − current environmental damage
      + net investment in real capital
      + net change in the value of human capital
      + net change in the value of the stock of natural capital

Furthermore, the net change in the value of the stock of natural capital should be computed as the value of the change in the stock of resources (capital gains should thus not be included) and where the value of the resource should be represented by a shadow price, reflecting the present value of the future marginal productivities of the resource. This construction differs from the recommendations from the UN, as it does not subtract "defensive expenditures," i.e., expenditures for protecting the environment and that it calculates the depreciation of nonrenewable resources as the change in the physical stock times a price and does not try to calculate the "Hicksian" income by introducing some arbitrary assumptions on future price and quantity changes.

Repetto, in two studies (1989) and (1991), has tried to estimate the importance of including the change in natural capital in the national accounts for Indonesia and Costa Rica. For Indonesia, Repetto showed that the average annual growth rate from the early seventies to the middle of the eighties would roughly halve if the depreciation of oil, forests and soils were taken into account.

7.4   Optimal growth

Given the NNP-index discussed in the previous section, could we expect that an economy would follow the optimal path? And if not, what are the issues involved? In the popular literature, a number of arguments have been raised against the prospects of markets doing the job. These do not only include the standard externality argument but quite many others such as that man only plans for his own lifetime and as it is finite, he does not look sufficiently into the future. In order to discuss this and other issues, we will in this section construct and study a very simple overlapping generation model.[3]

Consider an economy for a series of time periods $1, 2, \ldots, T$. In each time period there are two generations, one young, productive and one old and consuming. In the following period, the young generation has grown old and a new young generation is producing. In its planning, the young generation will take into consideration its consumption in the next period and accumulate wealth, the return on which can be used for consumption. Wealth can consist of both reproducible capital K and environmental resources Y. Thus we make the very important assumption that it is possible to assign property rights to the resource. The ultimate value of the environmental resources derives completely from their use as input in production. Here we differ from the discussion in the previous section where the stock of the resource not only provided inputs to production but were also valued directly by households. The current young generation will in this model have incentives to maintain the resource till next period, because it can then sell it to the new young generation and thereby earn a return in the form of capital gains. The new young generation will also look forward to the possibilities of selling the resource to the next young generation and so on. Thus, the present generation will implicitly through the future prices on the resource, given that it has rational expectations, take the importance of the resource for all future generations into account. Let us look into this in more detail.

Let the representative individual's life cycle utility be

$$U(C_t) + \frac{U(C_{t+1})}{(1 + \phi)} \tag{7-1}$$

where c_t is the consumption in his young period, and c_{t+1} is his consumption when he retires. $1 + \phi$ is his subjective utility discount factor. The budget constraint can be written

$$c_{1,t} + s_t + v_t^c = w_t \tag{7-2}$$

$$c_{2,t+1} = (1 + r_{t+1})s_t + v_{t+1}(1 + \mu)y_t^c \tag{7-3}$$

where s_t is the savings during the first part of life, $v_t y_t^c$ the expenditures on the purchase of the resource stock, w_t wage rate (we make the standard assumption and assume an exogenous supply of labour). In the second part of life, the individual can use the return on his savings $(1 + r_{t+1})s_t$ and the sales of the resource $v_{t+1}(1 + \mu)y_t^c$ for consumption $c_{2,t+1}$. Here we have assumed that the stock will grow spontaneously between the two time peri-

---

[3] This model is constructed along the presentation in Blanchard and Fisher (1989).

ods at the rate μ. If the resource is an exhaustible resource, $\mu = 0$. The assumption that the growth rate is constant is essential for some of the conclusions that will be drawn but probably not for the most important one which deals with missing asset markets.

Maximising utility subject to the budget constraint yields the following necessary conditions

$$u'(c_{1,t}) - (1 + \phi)^{-1}u'(c_{2,t+1})(1 + r_{1+t}) \leq 0 \tag{7-4}$$

$$u'(c_{1,t})v_t - (1 + \phi)^{-1}u'(c_{2,t+1})v_{t+1}(1 + \mu) \leq 0 \tag{7-5}$$

where equality holds when the corresponding variable in optimum is greater than zero. Assume for the moment that both conditions hold as an equality. Then it follows that

$$\frac{v_{t+1}}{v_t} = \frac{1 + r_{t+1}}{1 + \mu} \tag{7-6}$$

which is the Hotelling rule for a renewable resource or the basic arbitrage condition for intertemporal efficiency. If this rule holds, the individual is indifferent on the margin between investing in the resource asset or the reproducible capital asset. If $\mu = 0$, it becomes the original Hotelling's rule, and the allocation of the resource over time will be such that at least one condition for efficiency is satisfied, that is the price path will be efficient. However, this is not a guarantee for dynamic efficiency as we have seen in previous sections.

Let us assume that the production function is

$$Q = F(K_t, Z_t, L_t)$$

where K_t is the stock of reproducible capital, Z_t the input of resources and L_t the input of labour. We assume that F is linearly homogeneous so that we can express everything per unit of labour ($q = Q/L$, $k = K/L$, $z = Z/L$):

$$q_t = f(k_t, z_t) \tag{7-7}$$

Profit maximisation yields

$$f(k_t, z_t) - kf_k'(k_t, z_t) - zf_z'(k_t, z_t) = w_t \tag{7-8}$$

$$f_k'(k_t, z_t) = r_t \tag{7-9}$$

$$f_z'(k_t, z_t) = v_t \tag{7-10}$$

Let y_t be the stock of the resource per capita existing in the beginning of period t, i.e., $y_t = Y_t/L_t$. This stock is related to the desired stock held by the individuals by

$$y_t^c = y_t - z_t \tag{7-11}$$

where z_t is the amount used up as input in production per unit of labour input.

Conditions for market equilibrium for goods, capital, resources and labour can now be written. Because of Walras' law, it is sufficient to describe the equilibrium for capital, resources and labour. For capital, the present savings will equal the capital stock in the next period, which means that (n is the growth rate of labour)

$$k_{t+1} = \frac{s_t}{1 + n} \tag{7-12}$$

The equilibrium condition for the resource is

$$y_{t+1} = \frac{1 + \mu}{1 + n}(y_t - z_t) \tag{7-13}$$

This can be explained as follows. In the beginning of period t, there is a stock of resources Y. Part of this stock, Z_t, is sold to producers, and the remaining $Y_t - Z_t$ to consumers, i.e., $L_t y_t^c$. The resources used as inputs by firms are used up. The stock bought by consumers will be preserved and enhanced by the growth factor μ. Thus in the beginning of period $t + 1$ there will be a stock equal to $(1 + \mu)(Y_t - Z_t)$. Per unit of labour this means

$$\frac{1 + \mu}{1 + n}(y_t - z_t) \tag{7-14}$$

The equilibrium condition for labour is simply given by $L_t = 1$ (or any other exogenous value).

This set of equations can be seen as a dynamic system. We can solve the above equations for r_{t+1}, v_{t+1}, $c_{1,t}$, $c_{2,t+1}$, s_t, z_t, y_t, w_t as functions of k_t, y_t, r_t and v_t. All in all we then have a complete dynamic system consisting of four difference equations

$$k_{t+1} = \frac{s(k_t, y_t, r_t, v_t)}{1 + n} \tag{7-15}$$

$$y_{t+1} = \frac{1 + \mu}{1 + n}(y_t - z(k_t, y_t, r_t, v_t)) \tag{7-16}$$

$$r_{t+1} = r(k_t, y_t, r_t, v_t) \tag{7-17}$$

$$v_{t+1} = v(k_t, y_t, r_t, v_t) \tag{7-18}$$

Let us assume that this system has a unique steady state with a positive stock of the resource and a positive use of the resource as an input in production (in general there may not exist a steady state or there may be many, so the assumption is not innocuous as we shortly will see). It is easily seen that such a steady state is characterised by (steady state values are denoted by an asterisk)

$$r^* = \mu \tag{7-19}$$

that is, the interest rate and therefore the capital stock is determined by the growth rate of the resource. If that growth rate is zero, the decentralised equilibrium interest rate would also be zero, indicating that compensations to future generations should not be discounted. If there are many resources with different growth rates, the interest rate cannot be at the same time equal to all the growth rates which indicates that this model may be too simple to answer questions on allocation between resources (there would not exist a steady state). In spite of this, the result indicates that the decentralised equilibrium may reflect the future development of the resource use. The trade between generations will, in fact, make it profitable for the current generation to take the need of future generations into account. It can be seen that if the steady state exists and is unique, then it will also be a saddle point with the same properties as the saddle points we looked into in the previous section. Thus, if the present resource price is not too far from the "optimal" price, the economy will move to a neighbourhood of the steady state and stay for some time until it will deviate and either deplete the resource or accumulate the resource. Unless the economy is very far from an optimal path, the competitive prices will be approximately similar to the "optimal" prices and in particular the interest rate, as it is established on the markets, will correctly carry the information on the future productivity of capital.

The word optimal has not been defined in the discussion above. One way of looking at it is to consider a totally planned economy, in which the planner maximises

$$\sum_0^\infty (1 + R)^{-t-1}[u(C_{1,t}) + (1 + \phi)^{-1}u(C_{2,t+1})] + (1 + \phi)u(C_{2,0}) \tag{7-20}$$

where R is the planner's rate of time preference. If $R > 0$, this infinite series will converge and with $R = 0$, we interpret the maximisation as the overtaking criterion. The planner maximises this sum subject to the feasibility requirements

$$c_{1,t} = f(k_t, z_t) + k_t - (1 + n)k_{t+1} - (1 + n)^{-1}c_{2,t} \tag{7-21}$$

$$y_{t+1}(1 + n) = (y_t - z_t)(1 + \mu) \tag{7-22}$$

The necessary conditions for a maximum are

$$(1 + R)(1 + n)(1 + \phi)^{-1}u'(c_{2,t}) = u'(c_{1,t}) \tag{7-23}$$

$$f_k'(k_t, z_t) = (1 + R)(1 + n)u'(c_{1,t-1})/u'(c_{1,t}) \tag{7-24}$$

$$u'(c_{1,t})f_z'(k_t, z_t) - (1 + n)(1 + \mu)^{-1}u'(c_{1,t-1})f_z'(k_t, z_t) = 0 \tag{7-25}$$

Assuming the existence of a steady state in which a positive stock of the resource is preserved requires that

$$1 + R = \frac{1 + \mu}{1 + n} - 1 \tag{7-26}$$

As both the right hand side and the left hand side are exogenous, there is no direct reason why the equality should hold. However, it makes a lot of sense if the choice of a time preference rate is such that this equality holds. Then, it would reflect a kind of neutrality between generations. The faster the resource grows, i.e., the more abundant it is in the future, the more will the future be discounted, and the higher the growth of population, the less will the future be discounted. It now turns out that the steady state in the planned economy with this choice of interest rate coincides with the steady state in the decentralised economy. In particular, it follows that the real interest rate is once more equal to the growth of the environmental resource, μ. Thus, the decentralised economy will (at least in the steady state) perfectly reflect the needs of future generations in the current prices. With overlapping generations, selfish, shortsighted decisions taken by the current generation will through the equilibrium prices on resources and capital, in reality be farsighted and imply a neutral allocation between generations!

A necessary condition for the reasoning above is that $\mu > n$, otherwise the rate of time preference would be negative. The convergence of the infinite series is not obvious in case of a negative interest rate. If the resource is essential, a zero growth rate would imply a monotonic fall in consumption per capita, and if that decrease is faster than the decrease in the discount factor, the sum may converge, even with a negative interest rate. If the resource is not essential, i.e., its marginal productivity never increases without bounds, the resource may, with a low growth rate, become depleted. However, through the capital accumulation, the generations following that time period will be adequately compensated. In fact, after the time of depletion, the economy can be described by the standard model of overlapping generations. It is known that model does not predict dynamic efficiency and that there will in general be too high a stock of capital in the ensuing steady state. It is up to the future generations to reduce the capital stock in order to increase the steady state per capita consumption.

In any case, the market system will guarantee that the needs of the future generations will be reflected in the prices today.

The shortsightedness of the current generation is thus a nonissue in connection with sustainable development. Moreover, the interest rate and the prices on the resources will reflect (at least in the model) both the future value of the resource and the future opportunity of capital.

We have seen that under certain ideal assumptions, the market rate of interest will provide correct guidance in making decisions on intertemporal allocation. The proper criterion is to maximise the present value of the future stream of costs and benefits where this stream is discounted with the market rate. This holds true if there exists a set of complete markets for all goods and services and for all future dates or if the present individuals have perfect foresight and infinite horizon. If individuals have finite horizons because of their finite lives, the result still follows if they have perfect foresight. It has been argued above, however, that even if they have only short term perfect foresight (meaning knowing the rate of change of prices), the existence of overlapping generations will guarantee that the present generation will take into account the needs of the future generations and moreover, if the economy dynamically moves to a steady state, then the prices tend to reflect the true social costs, at least approximately. However, this is based on the assumption of the existence of a unique steady state. If such a state does not exist, then prices, and in particular the interest rate may be quite bad approximations of the correct or ideal prices.

Moreover, we have implicitly assumed above that all goods and services are bought and sold on markets. As is well known, this is not the case in the real world. Well defined property rights to the resources, which can be transferred between the generations simply do not exist for many of the environmental resources. For some resources, such as timber from forests, there exist such rights in some countries (but by far not in all). For most environmental resources it is not true, however because they are such that property rights cannot be defined. Clean air is shared by many people and it is obviously impossible to claim individual property rights to a piece of clean air. Thus, no one can buy a piece of clean air and sell it to the next generation.

In other cases, even if property rights can be assigned to the resource, it has for various reasons, not been done, so no one can secure for himself a piece of the resource and sell it to others. This is the problem of common property resources with free access.

Quite often, government interventions through taxes, pricing schemes, trade policies, land tenure rules, etc. create distortions in the incentive structures, with environmental degradation as a consequence. Because of these market and government failures, the need of future generations will not be reflected in the current prices and there is no reason why the current market interest rate should represent the ideal one. Only if it can be assumed that environmental resources that are not properly managed from a social point of view are of negligible value can the market rate be assumed to approximate the rate that would compensate the future generations.

Many resources are such that it seems improbable that a change in the management of these resources would affect the interest rate more than marginally. On the other hand, the global problems of climate change, the hole in the ozone layer and acid rain will in the long run probably mean quite a lot for the performance of the economy and therefore also for the market rate of interest. In many developing countries, where the resource base is rapidly being depleted, efficient management of environmental resources would also probably mean substantial effects on the interest rate.

The current situation is however one in which we do not have the analytical and empirical tools with which we can study the interrelations between interest rates and environmental resource management. Usually, the interest rate is assumed to be determined by some national guidelines, reflecting a combination of the opportunity cost of capital and the pure rate of time preference, when environmental resources are analysed economically. There is a strong need for the development of computable general equilibrium models which can be used to study the interactions between natural assets and the rest of the economy.

The immediate conclusions that can be drawn from the analysis in this paper are:

1. Future damage to environmental resources should be discounted at a rate that indicates the future productivity of all capital assets and not only human-made real capital.
2. Methods for predicting the future productivity of capital need to be developed.
3. Methods and techniques for assessing present and future values of damages to environmental resources need to be developed.
4. Methods and techniques for dealing with the uncertainty about future damage must be developed and principles for risk sharing between generations must be agreed upon. In particular, the theoretical knowledge that exists must be turned into operational tools.
5. Incentive systems must be designed in order to make decision makers take into account the consequences for future generations from their present decisions.
6. Institutions must be built that are responsible for undertaking the necessary investments in order to compensate the future generations from current resource degradation activities.

CHAPTER 8

The international regulation of biodiversity decline: optimal policy and evolutionary product

Timothy Swanson

8.1   Introduction

The problem of global biodiversity decline derives from the failure of states to consider the impacts upon global stocks when taking national resource regulation decisions. When stock depletion generates substantial externalities, as is the case with diverse biological resources, then decentralised (i.e., multinational) regulation of these global stocks will be necessarily inefficient. The biodiversity problem is, in effect, the predictable result of an imperfection in the existing global regulatory system with regard to diverse resources; decentralised decisions regarding diverse resources do not take into account the effect they are having on global stocks of biodiversity.

The statement of the solution to the global biodiversity problem is straightforward: decision-making in regard to diverse resources must take these stock externalities into consideration. However, this solution concept runs into the same difficulties as the underlying biodiversity problem, because it also must proceed through the same decentralised regulatory system. As natural resources are *national* resources, the mere recognition of these regulatory imperfections goes little further toward their solution. It is instead necessary to construct global incentive mechanisms which induce national regulatory actions consistent with the global object.

This chapter outlines the general nature of the global incentive mechanism required to internalise the value of biodiversity, but not until Section 8.6. First, it is necessary to provide a clear depiction of the nature of the problem to be solved; this description will then suggest the nature of its own solution. The first three sections of this paper undertake the task of formulating the nature of the regulatory problem that generates inefficient global biodiversity losses. In Section 8.2, the human-induced extinction process is described as the result of human choice in the conversion of the biosphere. In Section 8.3, the predictable failures in the decentralised regulation of this conversion process are described. Section 8.4 presents a model of the optimal policy problem regarding global biodiversity stocks. In Section 8.5, the costliness of these failures (i.e., the value of biodiversity as evolutionary product) is outlined. Then, finally, in Section 8.6, the general nature of the international incentive mechanisms necessary to redress these regulatory failures is discussed.

The international regulation of biodiversity depletion may sound like an unlikely or at least impractical project; Chapter 10 by Scott Barrett illustrates some of the difficulties inherent in this task. However, the point of this chapter is that the regulation of biodiversity is occurring right now, only on a wholly *national* basis and with a demonstrably suboptimal outcome. The international community needs to intervene, in certain specific ways, in order to redress the imperfections in the existing regulatory system.

8.2   The source of biodiversity decline: extinction as conversion

Prior to moving to a description of the regulatory problem, it is important to discuss the nature of the forces to be regulated. This is something I and others have discussed in detail elsewhere (Swanson 1993b, Swanson, 1995a); here, a brief survey of the forces at work will suffice. Extinction is not now an isolated, randomly occurring event applying to particular species; it is instead a systematically generated threat applying to nearly all species. It is a human-induced process that has been working across the face of the earth for about ten thousand years, the result of specific economic forces working at the global level – replacing naturally existing species (evolutionary product) with human-chosen ones. The regulation of biodiversity concerns the regulation of this process.

The decline of biodiversity may be conceived of as a *process of conversion,* the remaking of the naturally-endowed slate of species by human societies. That is, the extinction process, as it has occurred over the past ten thousand years, has been the direct result of human choices regarding the expansion of the ranges of certain heavily-utilised species. The impacts upon other species (range restrictions, overexploitation, population declines and extinction) have all been largely unintended consequences of these decisions. Nevertheless, these adverse impacts on the unchosen species are the direct result of human choices to expand their uses of others, in the context of strict resource (e.g., terrestrial, management) constraints (Swanson 1994).

Therefore, the modern problem of terrestrial extinction is a problem of the conversion of the range of life forms on earth to a smaller range of species, rather than a reduction in the biological productivity of the earth.

That is, the vast majority of biodiversity losses concern a process of conversion, rather than degradation. The extinctions that have been occurring, and the mass extinctions currently threatened, do not represent substantial reductions of global biomass stocks. Rather, the global biomass (in terms of net primary product) has remained relatively constant over this period (about 4% "lost"), while its constituency has been altered dramatically (with the prospect of the removal of 50% of all life forms) (Vitousek et al. 1986).

The "homogenising" character of the conversion process is the result of a dramatic change in relative investment rates regarding various species. About ten thousand years ago, the benefits from specialising in investments in certain species of animal and plant life forms became apparent. The results were the practices of domestication and cultivation that were fostered in respect to these species.

The specialised species, i.e., those which were domesticated or cultivated, began to attract a disproportionate amount of society's investments into biological productivity. These investments take the forms of: increased management expenditures on certain species (intensive cultivation and domestication); increased expenditures on capital goods compatible with these species (specialised capital goods); and, increased stocks of these specialised species (widespread utilisation).

In contrast, a wide and diverse range of life forms has attracted relatively little investment. The relative rates of investment result in conversions, directly or indirectly. This occurs most directly when a given territory of natural habitat is cleared and restocked with specialised species. It occurs more indirectly when the lack of investment in management (because these funds are attracted elsewhere) results in the overexploitation of the unmanaged habitat or species. In either case, the root cause is the absence of investment in this particular life form, resulting in its displacement by others.

Therefore, these differences in investment rates result in the conversion of terrestrial habitats from one slate of species to another. This is an extinction process based upon homogenisation rather than degradation. The modern day problem of extinctions need only be conceptualised as the by-product of the human-induced *global conversion process* (Swanson 1992b).

In addition, the global process of conversions has a dynamic force of its own. This force is an example of *hysteresis* or *system state economies,* i.e., a state within a system that provides incentives for its own replication (David 1985; Dixit 1992). It is the existence of sunk investment costs that creates this effect, because sunk costs cannot be retrieved on a different development path. This is precisely the nature of the global conversion process. The initial choices of the slate of specialised species were probably optimal in the local contexts in which they occurred (primarily the Near Eastern states); however, current conversions to these species owe far more to ten thousand years' worth of sunk investments than to their innate superiority. It is this long term inertia which maintains the force for conversion to this specific slate of species at the margin.

Therefore, the extinction process is a byproduct of a human-induced process of terrestrial conversions, now with a force of its own. The regulation of global biodiversity losses is the regulation of this global conversion process. The nature of the regulatory problem necessitates an analysis of the process, in order to determine whether the global conversion process has a "natural" stopping point, and whether this stopping point will be optimal from the global perspective. If not, then the pertinent issue is the nature of the international interventions required in order to halt the process at the globally optimal point.

8.3   Regulating the global biodiversity problem: past, present and future

For regulatory purposes, terrestrial natural resources are first and foremost national resources. This is the meaning of *the doctrine of national sovereignty* in international law. Each state has the unrestricted and exclusive right to determine (through act or omission) the management of the various natural resources that it "hosts." There is good reason in economics, as well as in law, for such a doctrine; it serves the role at the international level that property right institutions serve at the domestic. That is, under the doctrine of national sovereignty, there is a specific state that has the designated responsibility for the management of any given area of terrestrial resources. As with property rights, this is a useful mechanism for allocating management responsibilities to those agents best able to invest optimally in the resources.

However, just as with domestic property rights, problems can arise where environmental systems extend beyond institutional boundaries. This is the nature of *the problem of global biodiversity losses.* It is a problem resulting from the decentralised regulation of a process with clear global implications. In essence, the problem derives from the fact that the impacts of *national* resource exploitation on *global* stocks are not considered by individual regulator states. Therefore, the general problem of global biodiversity losses concerns the difficulty of adequately regulating global stocks of diverse resources in a decentralised (i.e., multinational) world.

Before moving to the issues of optimal international regulation, it is necessary to understand the nature of, and imperfections within, the existing, decentralised regulatory process. This section of this paper presents a specification of the existing global conversion process, its decentralised stopping point, and the reasons for its divergence from the global optimum. This constitutes a specific representation of the *global biodiversity problem*; it is a problem of a global process that is inadequately regulated in a decentralised (i.e., multinational) world.

The history of biosphere conversion: inefficient decentralised regulation of the global conversion process

This section tells a very short story about a very long period of time. Its purpose is to develop a schematic that is descriptive of the process by which the earth's biological resources have been converted. It encapsulates a "history" of the biosphere since the introduction of domesticated and cultivated species, some ten thousand years ago.

The history of the biosphere since that time may be best described as one of *sequential conversions: the alteration of the world's slate of natural resources on a state-by-state basis.* Conversion within a particular state has occurred as and when the idea of employing domesticated and cultivated varieties (and the methods of production that they imply) has arrived at that state at a particular point in time; then, the natural slate of resources is usually displaced by the domesticated. The sequential nature of this process is discernible in the nonuniform progress of the process across the world's states; some states are long-converted, while others have had little resource conversion within their borders. The conversion process has therefore diffused in general much more rapidly within particular states than it has between them. In order to simplify the exposition in this paper, it will be assumed that the conversion frontier moves between states, rather than within them, and that (on arrival of the frontier) the adoption of the ideas of modern agriculture (i.e., conversion to specialised varieties) is of the nature of an "all or nothing" proposition; this is the "sequential conversion process" analysed here.

Therefore, in this analysis, the time horizon relating to the biosphere is demarcated into two sections, representing the succession of states that have already converted their biological resources (the "past") and those states which have not (the "future"). The conversion frontier (the "present") resides at the *marginal state (MS)*: the state that is currently contemplating the conversion of its natural slate of biological resources.

The force that generates the progression through this series of states is the *global conversion process.* It is the perceived net benefit from the conversion of the state's naturally evolved slate of resources to the human-selected specialised slate (i.e., the slate of domesticated and cultivated varieties and their associated methods of production).

Over the course of the history of the biosphere, each state presented with the idea to date has undertaken the option of conversion, i.e., the net benefits to conversion have been perceived to be positive. With successive conversions, it is likely that the perceived net benefits to conversions will be declining, on account of both rising supply costliness as well as falling prices for the goods and services flowing from conversion. However, with so few states remaining within the nonconverted margin, it is apparent that the perceived net supply costs of conversions are not rising rapidly enough to deter conversions by themselves.

This process is depicted in Figure 8.1. Along the horizontal axis is the passage of time, representing sequential conversions of the diverse resources managed by individual states. The history of the evolved biosphere ends at the point of 100% conversion, when no conversion frontier remains on earth. The question addressed in Figure 8.1 is whether the global conversion process will continue through to totality (the elimination of the evolved biosphere), or whether there are forces that will halt this process.

Figure 8.1 depicts a scenario where there is little likelihood that the conversion frontier will be halted. First, it assumes that each state perceives a nonincreasing costliness to the conversion of its natural resources; that is, the costs of the specialised inputs (machinery, chemicals, and species) of converted production are nonincreasing to each successive state. This is represented by the downward sloping *supply of conversions curve* in Figure 8.1, which is the perceived marginal cost of conversion at each point in time (i.e., by each state).

Another reason why there is little prospect that conversion will halt is the lack of an adequate demand-related constraint on the process. The analysis within Figure 8.1 places no constraint on conversion arising out of a lack of substitutability between specific natural and specialised outputs, although the average benefit from additional units of identical outputs is declining. This is represented by a downward sloping linear *demand curve* (representing the perceived average benefit from conversion to specialised resources). The average benefit to conversion declines, because of the existence of some diverse products for which consumers hesitate to accept specialised substitutes. Then, as global conversion approaches totality, the marginal benefit to further conversions becomes very small due to consumer hesitancy to substitute (as forecast by John Krutilla in 1967), but never small enough (in Figure 8.1) to halt the process.

It is important to be precise concerning the nature of the "benefits to conversion," as they are only in the short run derived from the increased productivity of specialised resources. In the medium run (with other factors adjusting), the benefits from conversion are derived from very different sources.

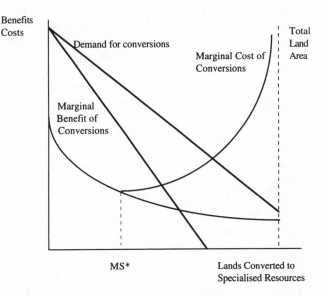

Figure 8.1. Optimal policy regarding conversions.

Even if it is believed that there is perfect substitutability amongst consumers between specialised and diverse resources, it is likely that the demand curve for conversions would be downward sloping. This is because the analysis in Figure 8.1 is of the longer run (the amount of time that will determine the quantity of evolved resources that will remain). In this timeframe, there will be a number of factors adjusting to the average productivity gains implied from land use conversions. In particular, the human population density is assumed to adjust in response to the increased average productivity of the land use. Then, in the assumed absence of effective population controls, the primary effect of the conversion process is the expansion of the human niche, i.e., the human population density expands within the converting state.

This means that, in the longer run, the benefits flowing from land conversions depend exclusively upon the social benefits to be derived from greater population densities. There is, of course, a long debate concerning the benefits to be derived from population growth. One side to the debate points to the historical evidence that population growth has been the catalyst of economic and social change over the past few centuries, with significant societal benefits (Simon 1992). The other side to the debate looks at the present and to the future, and finds little prospect for societal benefits in further population growth (Lee 1991).

The purpose of this analysis is not to take a position on one side of this issue or the other, but rather to explain the difference in perspectives that generates the two sides to the debate. If the average benefit curve to further conversions is downward sloping in the longer run (with populations adjusting), this implies that the strategy of high population density (for a given state) would initially afford substantial average benefits (in an otherwise low density world) – these result from the benefits to be received from the unique capabilities (urbanisation, industrialisation) that high density affords over low. However, as additional states convert to this strategy, there are two effects. First, it is likely that the average benefits from a strategy of human population expansion (urbanisation, industrialisation) decline with an increasing number of densely populated states. This is because the densely populated states are ready substitutes for one another in the population-based services (e.g., low cost labour-based manufacturing) that they can provide. Secondly, the *marginal benefits* of a high density strategy decline even faster than do the average, and these are the source of a pecuniary externality operating between the newly converted and the formerly converted states.

Therefore, in this framework, there is a germ of truth within both sides to the debate. Looking retrospectively, societies operating with high population densities (e.g., Europe and Japan) in a low density world have gained substantial advantages from the distinct opportunities that this factor has provided. However, with a large part of the world already converted, the average benefits from further conversions are now much reduced. Looking solely prospectively (from the vantage point of the marginal state forward), there is a much reduced benefit to individual land use conversions and the higher population densities they imply, even though in the past the benefits were much greater. The position that one takes in the debate simply depends on where within the global conversion process one stands when considering the average benefits to conversion (and hence high density).

This might indicate that the "population pessimists" have the upper hand in the debate within this model, since they seem to be focusing upon the more relevant notion of "average benefit," i.e., that average benefit which exists in the present and future. However, it is more likely that the pessimists are looking to the marginal than the average benefit to population growth, and deriving their conclusions from this. The marginal benefit to population growth (which is perceived by the converted but not the unconverted states) will fall much below the average benefit. This is attributable to a pecuniary externality, which drives the value of the uniform specialised outputs downwards with each successive entrant. A difference in perspective here (in terms of focusing on marginal versus average benefits) might explain the ironic nature of the current debate, with the pessimists within the relatively high-density states (of Europe and North America) preaching the benefits of population control to the relatively low-density states (of Africa and South America). One group might be pointing to the marginal benefits to such population expansions, while the other is reacting to the average.

In any event, all considerations indicate that the average benefit curve for further conversions (and hence population expansions) is downward sloping. However, if the demand for the flows from specialised species is not too inelastic (indicating that there is broad substitutability between species for basic needs such as food and clothing) and nondecreasing returns to scale apply to a handful of species over the entire global land area, it is possible that these conversions might continue through to totality, as depicted in Figure 8.1. Then the entirety of the output from the natural evolutionary process will be lost to the global conversion process.

The future of biosphere regulation: international institution-building for internalising externalities

There is another force which should enter into decision making concerning the conversion process. This is the geometrically increasing costliness of the final conversions of diverse resource stocks (or, alternatively, the marginal value of biologically diverse resources). This is represented by the upward sloping *marginal cost of conversions curve* in Figure 8.1, which includes the marginal opportunity cost of the loss of these diverse stocks of lands and resources.

This is an important externality within the decentralised regulatory process. At each point in time (i.e., with each successive conversion), this costliness is increasing but unaccounted for, because the opportunity costs of conversions in terms of lost global services are not included within the converting state's decision making framework. The "marginal state" considers only the perceived supply cost, not the actual cost of conversion.

The loss of diverse resource stocks necessarily entails global costliness, in terms of irreplaceable insurance and information services, and hence these losses should be considered as an opportunity cost in the supply of converted lands. The precise nature of these costs is described in Section 8.5. In fact, the marginal costliness of successive state conversions will be rapidly increasing, once a substantial part of the terrestrial surface has been converted to specialised resources. This is because the relationship between conversions of land area and loss of species stocks is nonlinear (MacArthur and Wilson 1967). Species-area functions are said to follow an Arrhenius (log-linear) relationship. Studies in island biogeography demonstrate just such an empirical relationship, indicating that the conversion of 90% of land area results in losses of about 50% of the species diversity. (Wilson 1988). The importance of this relationship lies in its implications regarding the final conversions of the residual territory. In general, this final 10% of conversions will entail as great of losses of diverse resources as did the initial 90%. Therefore, to the extent that global costliness is directly related to the loss of species diversity (as will be shown to be the case in Section 8.4), this costliness will be increasing geometrically with the final conversions.

Therefore, with the passage of time (and successive conversions), there are two countervailing forces which should determine the globally optimal stock of biodiversity: the benefits from specialisation and the benefits from diverse resources. The global conversion process should halt when the marginal global value of the next conversion is negative, or alternatively, when the marginal value of retaining the resources in an unconverted state is positive.

In the context of Figure 8.1, the global process of conversion should be halted by the force of the value of global biodiversity at the point of intersection between the marginal cost curve and the marginal benefit curve (at time t^* and state MS*), but this does not occur on account of the decentralised nature of the process. Individual states perceive the average rather than the marginal benefit from conversion, and they fail to internalize the increasing marginal cost of further conversions. An optimal international policy for biodiversity would then attempt to target this point through regulation of the global conversion process.

Therefore, an optimal international policy for biodiversity must halt the global conversion process at the optimal point in time (at the optimal "marginal state"), and it must do so (in a multinational world) through influencing the incentives perceived by those states. International institutions for biodiversity conservation should have as their objective the substitution of the "actual" for the current "perceived" cost curve within the decision making of marginal states. That is, international institution-building must take the form of investments in rechannelling the global values of diverse resources to the host states themselves. In this way, it is possible to achieve an optimal amount of conversion within a multinational world.

8.4   Optimal policy and the global biodiversity problem

This section makes concrete the various concepts developed within Section 8.3, in order to define the nature of an optimal biodiversity policy. Given the dynamics of the global conversion process (as described in Section 8.3), the problem may be defined as the choice of an optimal "stopping rule" with regard to this process. That is, optimal policy for biodiversity devolves in its simplest form to a determination of the *marginal state* at which the global conversion process is optimally halted. This section gives a model for this optimal policy, and describes the nature of the intervention required to implement that policy.

Optimal policy for biological diversity – a model

As developed in Section 8.3, the global biodiversity problem is best-described as a conflict between nationally rational and globally optimal decision making with regard to the global conversion process. This process may be conceived of as a *sequential bioeconomic process*. The passage of time in this model captures three important dimensions within this process: the movement of the "conversion frontier" to the borders of the next state; the change in the value of the remaining diverse resources (as global stocks erode); and the accumulation of information. The conflict arises on account of the predictable divergence between the local and global optima. The local "host state" will make a decision regarding the conversion of its diverse resources which does not consider the global benefits derived from these stocks, and this externality is the source of the global biodiversity problem.

Specifically, imagine that the states of the earth form two groups at any point in time, termed "North" and "South," the former being those which have previously converted their lands to specialised biological resources and the latter being those which have not. The issue (for purposes of optimal policy) is whether the marginal state (MS) in the group South should convert its biological resources when full global costliness is considered. The globally optimal stopping rule will halt the global conversion process at the point in time (i.e., the marginal state) when the global benefits to conversion are nonpositive.

In order to depict the optimal policy applicable to this process, imagine that, as the idea of biological specialisation diffuses across the earth, each individual state(s) determines whether to convert its naturally existing vector of biological resources (x_s) to a specialised slate of resources. Conversion consists of eliminating the state's land area dedicated to nonspecialised biological resources (R_s), and coincidentally the state's stocks (x_s) of diverse resources are also replaced by specialised stocks. The reduction in *global* stocks of diverse habitats (R) and resources (x) has impacts both on the individual, converting state (primarily in terms of changes in appropriable flows), and also on the welfare of all other states (primarily in terms of nonappropriable insurance and informational services).

The driving force behind such stock conversions is the nature of the flow generated by the different biological stocks. Specialised biological resources generate flows to their host state (y_s^{sp}), a vector of flows corresponding to a subset of species taken from a small menu available to all states (Y^{sp}). The important point about this vector is that it captures nearly the entirety of the flows of value emanating from that state's specialised stocks (x_s^{sp}). The degree of local appropriability of these flows is virtually one hundred percent.

On the other hand, a diverse stock of resources generates a vector of flows to the host state (y_s^d) which is very different from the specialised vector available through conversion. First, the dimensionality (in terms of species) of the individual state's "diverse resources vector" is several orders of magnitude greater than that of the "specialised resources vector" (millions of species as compared to dozens). Second, the rate of appropriability is far greater than it is for the diverse resource vector; that is, the appropriated flows (y_s^d) from diverse resource stocks (x_s^d) represent only a small proportion of the total flows from these stocks.

In addition, the global stock of diverse resources (x) generates a vector of flows (Y^d) which is again several orders of magnitude greater than the dimensionality of y_s^d in any one state; that is, there is little intersection between the diverse resource slates in the various unconverted states. In contrast, the global dimensionality of the menu of specialised species (Y^{sp}) is only in the dozens, and the Northern states have substantial commonality in flows.

The global conversion process can then be conceptualised as the sequential decision making by successive "marginal states" on whether to "change menus," from diverse to specialised resources. If conversion is elected, this is accomplished by means of noninvestment in diverse resource stocks (resulting in their removal by overexploitation and replacement), and contemporaneous investment in the specialised species (by selecting some subset from the global menu, Y^{sp}).

Diverse and specialised resources – definitions

$R_{(S \times 1)}$ ≡ the area of diverse habitat in each of the S states in group South

$x_{(S \times D)}$ ≡ the stocks of diverse resources in each of the S states in group South ("global stocks")

$Y^{sp}_{(N \times C)}$ ≡ the aggregate flows from global stocks of specialised resources

$Y^d_{(S \times D)}$ ≡ the aggregate flows from global stocks of diverse resources

$y_s^{sp}{}_{(1 \times c)}$ ≡ the appropriable flows from each of the c specialised species within (converted) state s

$y^d_{s(1 \times d)}$ $\equiv$ the appropriable flows from each of the d diverse species within (unconverted) state s

The optimal policy for biodiversity is the determination of the optimal stopping point in the global conversion process. The problem is to select the marginal state (MS) at which the conversion process should be halted, if the maximisation of aggregate global benefits is the objective.

The regulation of global biodiversity – optimal conversions

$$MS^{\Omega} \cdot \overset{\text{Max}}{\underset{MS}{}} E\left(\prod_G^{MS}\right) \equiv E\left(\sum_{s=1}^{MS-1} \prod_s(y_s^{sp};x^{ms},R^{ms})\right) \tag{8-1}$$
$$+ E\left(\sum_{s=MS}^{N+S} \prod_s(y_s^{d};x^{ms},R^{ms})\right)$$

where:

MS $\equiv$ the marginal state member of group South

$\prod_s$ $\equiv$ the present value of the flow of benefits to state s

x^{MS},R^{MS} $\equiv$ the global stocks of diverse resources and habitats given marginal state MS

$N+S$ $\equiv$ the number of sovereign states (decentralised management units)

$\prod_G^{MS}$ $\equiv$ the present value of the global benefits given marginal state MS

The two terms in Equation (8-1) represent the total benefits to the group North and the group South, respectively. The act of conversion shifts the marginal state in group South (MS) to the group North (implying that the second term then commences summation from $S = (MS + 1)$). This means that the marginal state has converted its biological assets to some subset of the specialised roster of resources.

This has the nationally perceived effect of shifting production to the (appropriable) flows from specialised rather than diverse species (i.e., from y_s^d to y_s^{sp}). It also has a global impact, in terms of a reduction in the global stocks of diverse resources (from x^{MS}, R^{MS} to x^{MS+1}, R^{MS+1}). Since global stocks of diverse resources contribute to production in both sectors, while only a portion of this flow is locally appropriable (in the form of y_s^d), the globally optimal stopping rule in the conversion process must take into account this stock externality to the Northern states.

Therefore, globally optimal regulation of biological diversity takes the form of halting land use conversions, even when these are perceived to be in the national interest, if they are contrary to the global interest. Because the conversion process has occurred historically on a state-by-state basis (rather than uniformly across all states), the globally optimal stopping rule also has the effect of determining the final size and constituencies of the groups "North" and "South." That is, it creates a "division of labour" in world biological production, with one group of states specialising in the production of diverse resource flows and the other specialising in the production of specialised resources.

The nature of optimal intervention

It is important to be precise about the meaning of the vector y_s^d – the flow of goods and services which come within the value function of the host state – as it is this definition which captures the essence of the necessary intervention. The vector y_s^d *is best conceptualised as that part of the flow from diverse stocks which is appropriable by the host state under existing institutions.*

Some of the benefits of diverse resources are flows, but flows whose appropriation by the host state (via unilateral action) is comparatively costly (or impossible). To the extent that these nonappropriable flows constitute a significant portion of the benefits from diverse resource stocks, there is an inefficient bias against holding these forms of biological assets.

As states continue to change the form of their holdings of biological assets, always choosing from the same small slate of specialised resources, the biosphere on earth assumes a very particular form. There are greater and greater numbers of a very small subset of species on the earth (as the group North grows) and there are fewer and fewer numbers of all other species (as the group South shrinks). Given the role of diverse resources (shown in the next section), this implies that the nonappropriable values of diversity will be increasing with successive conversions.

This determines the nature of the intervention necessary for the optimal regulation of extinction. The objective is to install an optimal stopping rule to the global conversion process, so that this process is halted when the marginal values of diverse and specialised resources are in balance. One mechanism which will implement such a solution is to bring more of the nonappropriable stock-related values within the host state's decision making process, i.e., to internalise the global externalities of diverse resource stocks. In terms of the model outlined above, this is accomplished by means of shifting more of the flow from diverse resource stocks (x_s) into the appropriable flows vector (y_s^d). The impact of internalisation will be to cause marginal states to perceive the level of opportunity costs implicit within the marginal cost curve in Figure 8.1, inducing globally optimal decision making by individual states. This has the effect of implementing a stopping rule at the globally optimal point in the conversion process.

This is the probable nature of the required solution to the global biodiversity problem for two reasons: first, since diverse resources all lie within sovereign states, the mode of intervention must operate through the decision making frameworks of the host states; secondly, so long as all states have a right to development, this solution concept protects that right in a way consistent with the protection of global values. In short, focusing upon enhancing the appropriability of global externalities affords these states an alternative path to development, rather than denying "Southern" states development in order to maintain the global optimum.

Therefore, the general nature of the global biodiversity problem is the presence of significant levels of nonappropriable benefits from the maintenance of diverse resource stocks. The general nature of the solution is the internalisation of these same benefits within the decision making process of the marginal states.

Clearly, the existence of such stock-related (i.e., nonappropriable) flows is a necessary condition for the existence of either a problem of biodiversity, or a solution. The existence of these benefits, and their nature, is the subject of the next section. In Section 8.6, the paper returns to the subject of the optimal form of international intervention.

8.5   The values of global biological diversity

The value of diverse biological resources is, in terms of the model of the previous section, equal to the total flows from all of the diverse resource stocks that exist on earth (x). This is the value that should be given effect within the context of an optimal regulatory policy, in order to halt the global conversion process at the globally optimal point. This value is defined in the next section as the "marginal value of biological diversity (MVBD)." It is a functional notion of value alone, devised because it is required for the purpose of defining the regulatory objective. This concept is the subject of the first part of this section.

Before proceeding to a discussion of this concept, it is important to note the distinction between the values of biodiversity and diverse biological resources, as the two are not synonymous. The latter is an all-inclusive category, encompassing the tangible and intangible flows from all biological resources that exist in areas that have not been subject to human conversion. In contrast, the former term corresponds only to the value of "diversity" (as opposed to "uniformity" or "homogeneity"); it is the value that flows from the mere fact of nonconversion.

In this sense *biodiversity* represents two distinct components: first, it is the value of the goods and services generated by the evolutionary process (as they encapsulate a 4.5 billion year history of adaptation and coevolution); and second, it is the value of retaining a production strategy on earth distinct from the "specialised" one. These values, i.e., the values of global biodiversity, are the subjects of the latter three subsections of this section.

The marginal value of biodiversity (MVBD)

A very concrete meaning may be given to the concept of *the total value of global biological diversity*. This is the opportunity cost of the conversion of diverse resources to specialised ones. There are both "total" and "marginal" concepts to be distinguished. The total value of global biological diversity would correspond to the total area under the marginal cost curve in Figure 8.1. *The marginal value of biological diversity (MVBD)* would correspond to a specific point on this curve, i.e., the costliness of taking the marginal step in the conversion process.

In both cases the concept corresponds to the opportunity cost of further conversions. However, the distinction is important because one is an operational concept and the other is not. It will never be possible to give precise meaning to the concept of the total economic value of biological diversity. This is because the diversity of life forms on earth is one of the fundamental components for maintaining stability in the biological production system sustaining human societies. As diversity goes to zero (total global conversion), the level of instability introduced implies that there is little prospect for human production systems to sustain themselves over any significant time horizon. Therefore, as conversion spreads to the last corners of the earth, its opportunity cost must be unbounded, as all human-sourced values must depend upon the maintenance of the biological production systems that sustain human societies. This is represented in Figure 8.1 by the area under the marginal cost curve, which is unbounded as conversions approach totality.

The functional notion of economic value as applied to biological diversity is the marginal one. The value of biological diversity cannot be divorced from the sequential decision making process of which it is an essential part. *MVBD* is the global opportunity cost of the marginal state switching from group South (the unconverted suppliers of diversity services) to group North (the previously converted states). It is also the force which must be effected if the global conversion process is to be brought to a conclusion short of total conversion.

$$\text{MVBD} \equiv \prod_G^{MS} - \prod_G^{MS+1} \tag{8-2}$$

given: $\prod_G^{MS} \equiv \sum_{s=1}^{(MS-1)} \prod_s (y_s^{sp}; x^{MS}, R^{MS}) + \sum_{s=MS}^{N+S} \prod_s (y_s^d; x^{MS}, R^{MS})$

where:

$\prod_G^{MS}$ $\equiv$ the present value of global benefits with marginal state MS

$x^{MS}, R^{MS} \equiv$ the global stocks of diverse resources (lands) with marginal state MS

$N + S$ $\equiv$ the number of states regulating terrestrial production

Equation (8-2) states that the MVBD, at a given point in the global conversion process, is the difference between the global present value of all biological production given existing stocks of diverse resources and the global present value of all biological production given the reduction in such stocks occasioned by the transfer of the marginal state between groups, i.e., the conversion of the marginal state's diverse resources.

This concept further demonstrates the nature of the global problem. It is a problem within a process, whereby continuing conversions occur not to increase aggregate overall value, but rather to increase aggregate appropriable value. There is little interest in domestic investment in supporting global values that are not channelled through the supplier state. Therefore, another means of representing the MVBD is in regard to its appropriable and nonappropriable components. Equation (8-3) restates Equation (8-2), but it now gives the change in the flow of values in two parts: those that are appropriable by the host state (8-3A) and those that are flowing generally to the global community (8-3B).

Appropriable and nonappropriable components of MVBD:

$$\text{MVBD} \equiv (A) \frac{\Delta\Pi_G}{\Delta y_s^d} - \frac{\Delta\Pi_G}{\Delta y_s^{sp}} + (B) \frac{\Delta\Pi_G}{\Delta x_s} + \frac{\Delta\Pi_G}{\Delta R_s} \qquad (8\text{-}3)$$

Equation (8-3) gives another general rendition of the global biodiversity problem; that is, it restates the reason why the local and global optima diverge. This divergence is attributable to the fact that the global conversion process is currently regulated by the force encapsulated by the first term in Equation (8-3), while the globally optimal regulation would incorporate the sum of the two terms in that equation.

This is because Equation (8-3A) represents the appropriable component of MVBD; it is the part of the change in flows of value that is channelled through the host state. This part of MVBD is generally negative, as states continue to convert their resources for the greater local productivity gains achievable through conversion to specialised resources. The potential force for halting the conversion process derives from the values encapsulated within Term (8-3B): the nonappropriable flows from the diverse resource stocks. The value of this term is demonstrably positive, and when it comes to outweigh the value of the appropriable elements [Equation (8-3A)], the process should be halted (from the global perspective). However, given nonappropriability of these values and decentralised decision making, there is a divergence between the local and global optima.

The remainder of this section discusses the nature of the "stock-related" (i.e., *nonappropriable*) benefits that are encapsulated within Equation (8-3B). These are the true "global values of biological diversity," i.e., the values that flow uniformly to the global community from the mere fact of nonconversion. In the ensuing discussion, the stock-related values of biological diversity [Equation (8-3B) above] are broken down into three distinct components corresponding to their static (*portfolio*) *value* and also to their dynamic value (i.e., the *expected value of information in the context of retained options*). In each case the value is developed in the context of the sequential decision making model discussed above. That is, it is the nonappropriable element of MVBD (i.e., the global value of the marginal state's decision not to convert its resources) that is being analysed.

The portfolio effect – the static value of biodiversity

Once biological diversity is considered outside of a deterministic framework, there is one obvious advantage which diverse resources have over specialised, i.e. their "pooling" capacity. If the global community is concerned not only with the mean yield of its biological resources but also with its variability, then their capacity to reduce global variability (via the pooling of distinct assets) is a desirable trait.

Global conversion upon a small slate of specialised species necessarily increases the variability in global yields. This is because the aggregate variability of all biological asset yields is not the simple summation of the individual variabilities of these assets. Aggregate variability instead depends crucially upon the independence of asset yields, i.e., the absence of a systematic correlation between them. Even if each of the different forms of biological assets has the same innate periodic variability (σ^2), the aggregate variability of these assets is equal to that variance divided by the number of independent assets (e.g., σ^2/C). This is known as the *portfolio effect,* and it derives from the fact that independent variabilities will have a cancelling out effect within the portfolio (Dasgupta and Heal 1979).

Consider again the sequential decision making model introduced earlier. In this model the marginal state must choose between biological production methods with a diverse roster of outputs (y_s^d) and specialised biological production with a small slate of outputs (y_s^{sp}). Here it will be assumed for simplicity of exposition that all members of the group North produce the same vector of specialised resources (of dimensionality c). Members of the group South each produce a much wider range of outputs (each of dimensionality d) chosen from a much wider range of possibilities (of aggregate dimensionality D). The key difference is the dimensionality of the two vectors. In matter of fact, states in group North choose their individual production vector from an aggregate vector Y^{sp}, whose dimensionality (C) is measured in the dozens. The important difference is that the dimension d is probably in the hundreds or thousands, and the dimension D is certainly in the millions, in contrast to the dimensions c and C which are only in the dozens at most.

The difference in dimensionality is important for the creation of a "portfolio effect" both locally and globally. Any state that retains its diverse resource vector rather than converting to the specialised maintains a much broader range of assets upon which to rely (d rather than c). Equally, this state is also investing in maintaining the existing dimensionality of the aggregate diverse resource vector (i.e., D). This contributes to a portfolio effect at the global level.

This may be demonstrated in the context of the framework developed above. First, it is important to note that each of the components of the production vectors (Y^d) and (Y^{sp}) is, at base, the same commodity: the flow of biomass from stocks of base resources (land). This flow might be measured in a common unit (e.g., usable mass, volume or energy), and then these vectors would simply represent the usable flows of energy/matter derived from various forms of production. In this schematic, a "species" is simply a distinct "method of production" for a common production unit, i.e., usable biomass.

In fact, this is a biologically sound description of the various components of these vectors; a species does represent a distinct method (or life form) through which the sun's energy is channelled through to human societies. The "portfolio" value to retaining a diversity of species thus derives from the maintenance of a range of different methods for channelling the sun's energy to human society; alternatively, the cost of successive conversions is the increasing reliance of an ever larger human society (built upon the increased average productivity of specialisation) upon a decreasing number of available methods of production.

To demonstrate this effect and the costliness it implies, assume that the annual production of usable biomass/energy is achieved by means of the use of the range of existing species. Global product is determined by the sum of the global yields for each of the components of Y^d and Y^{sp}.

That is, these aggregate production vectors may now be considered to represent D and C observations respectively on a common unit of production, i.e., the aggregate biomass/energy output from $D + C$ distinct methods of production. Given this definition, the annual productivities of all species will be a random variable (e.g., defined in terms of energy yield per unit of base resource), which may be described by reference to the various moments of its distribution in a given year.

Also, assume that there is some "global welfare function," $W(Y^d + Y^{sp})$, which captures the global community's willingness to trade-off between mean biological productivity and its variability. That is, it will be assumed that $W(Y)$ is a well-defined function over at least the first two moments of Y^d and Y^{sp}. Then this function merely expresses the notion that the global community is unwilling to pursue higher mean yields blindly, i.e., there must be some consideration of the higher moments of the expected flow of benefits. It is then possible to capture the nature of this trade-off, between mean production and its variability, in the following expression of the expected marginal value of biological diversity (via application of Taylor's expansion around the mean production levels).

MVBD – impact of conversion on mean and variability

$$E[\text{MVBD}] \equiv E\left(\prod_G^{MS}\right) - E\left(\prod_G^{MS+1}\right) \tag{8-4}$$

$$E[\text{MVBD}] \cong [W(Y_s^e) - W(Y_s^{sp})] + \tag{8-5}$$

$$\frac{W''}{2}\left(\frac{\sigma_{y^d}^2}{2} + \text{cov}(Y^d, y^d) + \text{cov}(Y^{sp}, y^d) - \text{cov}(Y^d, y^{sp}) - \text{cov}(Y^{sp}, y^s) - \frac{\sigma_{y}^{2sp}}{c}\right)$$

where:

d(or c) $\equiv$ the number of diverse (specialised) species in the marginal state MS

$W(\cdot)$ $\equiv$ the "global welfare function"

$\sigma_{y^{sp}}^2$ $\equiv$ the variability of biological production in MS with specialised species

$\sigma_{y^d}^2$ $\equiv$ the variability of biological production in MS with diverse species

The first terms of Equation (8-5) (i.e., the portion within the first brackets) correspond to the appropriable flows from conversion [as in Term Equation (8-3A)]. This once again captures the nature of the force for conversion, i.e., the higher expected mean yields from conversions to specialised biological assets. It is because this part of the expected value of diverse resources is negative that marginal states continue to undertake conversions.

The second term in Equation (8-5) (i.e., the portion within the second brackets) captures the global impact of the marginal conversion in terms of global yield variability; the source of the portfolio effect. This impact on MVBD is invariably positive, i.e., marginal conversions increase global variability and thus reduce global welfare.

Equation (8-5) may be broken down into three distinct parts. First, there is a species-specific portfolio effect from the use of a nonspecialised species. A specialised species is human-selected for certain traits, and then much of all future generations derive from this single selection, eliminating much internal genetic variability. A diverse biological resource is itself a wider portfolio for production. Second, there is a nationwide portfolio effect from retaining diverse resources. That is, domestic variability of production is reduced because the state retains a larger number of independent methods of production. Thirdly, there is a distinct international portfolio effect. The retention of any single state's diverse resources has the potential to reduce the aggregate covariance between national products.

In sum, the marginal value of biological diversity with regard to the value of diverse resources in reducing global biological yield variability is necessarily positive. The conditions that determine this are as follows:

MVBD – portfolio effect (PE)

$$E[\text{MVBD}]_{PE} \cong \frac{W''}{2}\left(\frac{\sigma_{y^d}^2}{d} + \text{cov}(Y^d, y^d) - \text{cov}(Y^{sp}, y^{sp}) - \frac{\sigma_{y^{sp}}^2}{c}\right) > 0 \qquad (8\text{-}6)$$

since:

(A) $W'' < 0$ (with risk aversion)
(B) $\sigma_{y^{sp}}^2 > \sigma_{y^d}^2$ (species-specific *PE*)
(C) $d > c$ (intranational *PE*)
(D) $\text{cov}(y^{sp}, Y^{sp}) > \text{cov}(y^d, Y^d)$ (international PE)

These terms give precise meaning to the various components of the global portfolio effect, as described.

First, Term (8-6B) represents the species-specific portfolio effect, in that a diverse species contains a wider range of genetic variability than does a specialised. For each diverse species, the inherent variability of production is reduced by reason of this innate portfolio of distinct characteristics.

Term (8-6C) captures the idea of greater variability stemming from the use of a smaller slate of resources within the marginal state – the intranational portfolio effect. Assuming that each species represents an "independent" method of production, in that the production from each is not correlated with that of other distinct species within that state, a wider menu of resources will reduce intrastate variability.

Term (8-6D) captures the nature of the interaction between production methods in use in the various states. In group North, the aggregate variability of the group must necessarily be greater when productive assets are more closely correlated, because there is less opportunity for production effects to "net out." This implies that the covariance between member states in group North will be higher than the covariance between member states in group South. Therefore, the marginal state's decision to switch between the two groups will necessarily increase the covariance of global yields – a reduction in the international portfolio effect.

Therefore, the movement of the marginal state from group South to group North represents an unambiguous increase in global yield variability, precisely because this conversion represents the movement on a global basis towards universally more specialised production. To the extent that global variability matters to human societies, then the increasing first moment of biological yields must be set off against the also increasing second moment of these same yields. It is important to emphasise that this is not a form of variability that can be resolved through other forms of insurance contracting, as this is increasing *global variability*. Reductions in the portfolio of diverse biological assets clearly have this effect on a global basis.

Dynamics: the value of exogenous information

The use of this sequential decision making model of the global conversion process highlights the importance of time and uncertainty. One of the meaningful facets of the passage of time is the accumulation of information, in the sense that an uncertain outcome is revealed in a subsequent period. A path must be chosen in the first period with only probabilistic beliefs as to resulting positions in the later periods, while the actual decision making framework (or "state of nature") applicable "now" will only be revealed in those later periods. In decision theory, information accumulates over time in the sense that outcomes of random variables affecting the decision maker's framework are revealed, and beliefs as to the future are better defined (Cyert and DeGroot 1987).

Therefore, placing the problem of global biodiversity into a sequential decision making framework also places the role of information accumulation at the core of biodiversity. In a dynamic framework, halting the conversion process equates with purchasing time and information. This makes the expected value of this information one of the fundamental forces for halting the conversion process.

Sequential decision making regarding resource conversions implies the passage of time, and one component of time is the accumulation of information. That is, uncertainties are resolved with the passage of time. Since the passage of time in this model is linked to decision making regarding the conversion of the marginal state's biological resources, one clear trade-off is the postponement of the marginal conversion in order to acquire the period's information.

MVBD – exogenous information (XI)

$$E[\text{MVBD}]_{XI} \equiv \left(\frac{\partial E\left(\prod_G^{MS}\right)}{\Delta I} - \frac{\Delta E\left(\prod_G^{(MS+1)}\right)}{\Delta I} \right) \frac{\Delta I}{\Delta t} \qquad (8\text{-}7)$$

where:

$\dfrac{\Delta I}{\Delta t} \equiv$ the addition to the information set over time

Equation (8-7) incorporates one definition of time; i.e., the process by which relevant information arrives at the decision maker. Then, information may be seen as eliminating relevant uncertainties with its arrival.

This value of biological diversity is unambiguously positive if two conditions are met: (1) information relevant to decision-making does in fact arrive by reason of an exogenous process over time; and (2) the conversion of the marginal state's resources reduces the dimensionality of the gross biodiversity vector $\{\mathbf{Y}\}$. These conditions guarantee nonnegative value because an irreversible narrowing of the choice set over time (in terms of reductions in the dimensionality of the gross biodiversity vector) renders information useless which would be otherwise valuable in the decision making process. Information is valuable, but only if the choices that it implies remain available.

The value of exogenous information (XI) – the option value of biodiversity

$$E[\text{MVBD}]^{XI} \geq 0 \qquad (8\text{-}8)$$

if:

$$(8\text{-}3A) \quad \frac{\Delta I}{\Delta t} > 0$$

$$(8\text{-}3B) \quad Y^{MS+1} \subset Y^{MS}$$

where

$Y^{MS} \equiv$ the gross set of global biological diversity with marginal (unconverted) state MS

This makes clear what the concept of *option value* means in the context of biological diversity. It represents the value of retaining the larger choice set until the next period's information arrives. It is, strictly speaking, the *value of flexibility* in sequential decision making, or *the expected value of information* (Hanneman 1989). Its value in this context is clearly positive.

This result differs from much of the literature on option value, which is inconclusive as to the sign of option value (Johansson 1987). The unambiguity of the result in Equation (8-9) is derived from two differences in the analysis.

First, this analysis focuses on a global process represented as a sequence of restrictions of the global choice set. That is, this model presents the problem of global biodiversity as a sequential narrowing of the choice set with regard to the methods of production ("species") available for capturing biological product. If two distinct sets have an equal number of different elements (or two production vectors have common dimensionality but distinct components), then "option value" (the value of retaining one rather than the other) is indeterminate *ex ante*. However, the problem of global biodiversity is best represented as a narrowing of the entire global choice set (of the available methods of production) rather than a substitution between elements. Then, the fundamental reason for indeterminacy is removed.

The primary reason that option value is clearly positive in this analysis is attributable to the specificity with which that term is used here. Here the "option value" of biological diversity refers only to the dynamic values flowing from *diversity,* and this excludes many of the other values of diverse biological resources. If these other values were included, then the marginal value of nonconversion might very well be negative.

That is, "option value" as used in the literature is sometimes equated with a concept analogous to the *marginal value of biological diversity,* as that term is defined in Equation (8-2). However, MVBD may be subdivided into two further components, the appropriable and nonappropriable flows indicated in Equation (8-3), and option value (as used here) concerns only a part of those nonappropriable flows found in Equation (8-3B). Specifically, option value as a concept most appropriately applies only to those values that flow from the existence of a dynamic facet to a problem, and the uncertainty inherent in sequential decision making. (Miller and Lad 1984; Pindyck 1991).

Therefore, the change in global values from marginal conversions may be segregated into three distinct categories: (1) the difference in the value of the appropriable flows from biological resources; (2) the static nonappropriable (i.e., portfolio) value of diversity; and, (3) the dynamic nonappropriable (i.e., expected value of information) value of diversity. Many times, the term "option value" has been applied to the aggregation of some combination of these three effects (yielding ambiguous results), but here it is used only in regard to the values derived by reason of the use of a dynamic decision-making framework. That is, option value concerns only the third category of diverse resource values listed above.

Option value is thus restricted here to mean only the value to be derived from retaining flexibility in a sequential decision making framework. With the arrival of information over time, the value of this retained flexibility must be positive. In this sense, the option value of biological diversity must always be positive.

It is logical to use this more restrictive definition in regard to the option value of biodiversity. The first two categories of value listed above are static values, i.e., they would flow from decision making even in the context of a one-period framework. On the other hand, the expected value of information (or "option value") applies only if there is a future. The value of such information will ultimately flow to the global community in terms of either increased mean yields or reduced variability; however, retaining these options has value even in earlier periods, to the extent that it is expected that their retention will be important. Diversity is retained in order to provide a dynamic as well as a static form of insurance.

It remains to explain why it is that information will necessarily arrive through an exogenous process, rendering the retention of these options valuable. In sequential decision-making, information is the occurrence of nondeterministic change in the decision making environment. That is, between periods there must be some relevant alteration in the environment that cannot be predicted with certainty; the passage between decision periods reveals the "state of nature" which could otherwise only be probabilistically projected.

The very nature of the biological world assures precisely this result. It is the very essence of a dynamic system, in which the processes of mutation, selection and dispersal continuously alter the natural "state of nature." In regard to small organisms, such as bacteria, viruses, and insects, these biological processes can occur very rapidly, literally reproducing thousands of generations in a single year. The biological process is evolutionary, not deterministic, and to the extent that it can be understood, it is too complex to predict.

It is the continuous state of motion within the biological world which guarantees that time produces relevant information. The nature of this information in a biological world is the type and extent of the shifting of the human niche. The insurance that we have for adapting to such shifts is the diversity of the species upon which we rely, or upon which we might rely. Marginal conversions represent losses of such options. The expected marginal value of biological diversity includes as a component the expected value of receiving information prior to the foreclosure of options. The value of this component is clearly positive.

Dynamics: the value of endogenous information

It was noted above that the intrastate variability of a larger vector of resources was likely to be smaller, and that states in the group South were the more likely states to rely upon a larger vector of resources. In fact, it is also likely that these states will also use biological assets with greater inherent variability in yield (and value). This is because, assuming regular concave individual utility functions, specialised production probably will be directed toward high mean/low variability biological assets. The variability occurs in the aggregate, because so many individuals specialise in precisely the same assets. In fact, from many years of use, the individual distributions of the yield from specialised biological assets will be very well-known, implying a very small amount of information to be gained from their use.

Unused, and less used, biological assets are relatively unknown quantities. The expected variability of any individual nonspecialised biological asset is quite large, especially relative to its expected mean, precisely because so little is known about it. The expected information acquisition from the exploration of these commodities is much higher than it is with the specialised assets, and this *exploration value* is positive.

Therefore, there is a differential informational value to the exploration of the nonspecialised biological assets that derives directly from the fact of their relative obscurity. This is termed the *expected value of endogenously generated information*.

MVBD: endogenous information (NI)

$$E[\text{MVBD}]_{NI} \equiv \left(\frac{\Delta \prod\limits_{G}^{MS}}{\Delta I} - \frac{\Delta \prod\limits_{G}^{MS+1}}{\Delta I} \right) \frac{\Delta I}{\Delta y} \frac{\Delta y}{\Delta R} \frac{\Delta R}{\Delta t} \tag{8-9}$$

Equation (8-9) states that nonconversion not only maintains the choice set at its differentially greater size, it also maintains the information flow at its differentially greater rate. The marginal conversion reduces the potential information set and the potential choice set in a single act. The reduction of either alone is costly in a sequential decision making framework. Therefore, the expected value of endogenous information is also invariably positive, being a mere subset (but a very important subset) of the entire information flow occurring between decision points.

Conclusion: The Values of Global Biological Diversity

The value of global biological diversity has been given a very specific definition in this section. It is the global impact of the conversion of the marginal state to specialised biological resources. This global impact has four components, two static and two dynamic (i.e., information-based). The information-based values of biodiversity will ultimately feed through the static components; however, in any given period there is also the expected value of this future flexibility.

The components of the value of global biological diversity

$$E[\text{MVBD}] \equiv (A) \frac{\partial \Pi_G}{\partial (y^d - y^{sp})} + (B)\ PE + (C)\ E[XI] + (D)\ E[NI] \quad (8\text{-}10)$$

As discussed in this section, the sum of the values of terms (B), (C) and (D) in Equation (8-10) is strictly positive in the context of the marginal state's decision whether or not to convert its diverse resources. However, it is also clear that the value of term (A) has been pronouncedly negative over many years, inducing successive states to undertake resource conversions.

It is not really possible to say anything about the relative magnitudes of these two forces. The force for conversion, term (A), flows through a vector of outputs which are clearly appropriable by the host state. The values of biological diversity, on the other hand, are clearly nonappropriable values. Under existing institutions, much of the value of global variability reductions is treated as a pure public good. For example, agricultural gene banks are created as "open access institutions," disallowing any one state's appropriation of the benefits of a resource treated as the common "heritage of mankind."

Therefore, in comparing these two forces, one for and one against conversion, their most distinguishing feature is their degree of appropriability. Under existing institutions, it is clearly not possible to gauge the relative magnitudes of these two forces. However, this does not imply that there is no value to biodiversity, only that it is not being given effect.

8.6   The international regulation of biodiversity decline: suggested institutions

The nature of the global biodiversity problem is one of imperfections in the decentralised (i.e., multinational) regulation of the conversion process, because individual states do not take into consideration the consequences of their conversion activities on global stocks of diverse resources. The obvious solution to the global biodiversity problem is to halt the global conversion process at its global optimum, taking into consideration these stock-related externalities (Swanson 1992a).

This "solution" is also a regulation problem, as it must also be implemented through the decentralised process that created the biodiversity problem. The agents with control over the conversion process are the individual host states, under the doctrine of national sovereignty, and the global community must take action regarding conversions through these agents. Therefore, the nature of international regulation regarding conversion is necessarily the construction of *international incentive mechanisms,* devised to induce individual states to regulate conversion in accord with global objectives.

The fundamental purpose of these incentive mechanisms must be to bring the global effects of diverse resource conversions within the decision making framework of the marginal states. That is, as discussed in Section 8.3, the object is to internalise the externality, causing individual states to recognise the true opportunity costs of diverse resource conversions. In essence, the incentive mechanism should have the effect of bringing the MVBD (the marginal cost curve in Figure 8.1) within the decision making framework of the marginal states.

The remainder of this section considers the general nature of this intervention, where the assumption is that the object of the global community is to implement indirectly the optimal stopping rule that cannot be implemented directly. That is, the objective is to halt conversions at the global optimum, given the globalised values of biodiversity, in a world in which there is little information on the latter and no inherent power to do the former.

Investing in the internalisation of biodiversity values

The optimal policy for regulating the conversion decisions of individual owner-states is the creation of systems that cause these states to consider the external effects of their decisions. In economic terms, the objective is *the internalisation of the global stock effects* of diverse resources in the owner-state's decision making framework. The rationale is that, if an owner-state considers the global benefits rendered by diverse resources when making its conversion decision, then it will only decide in favour of conversion when that is globally optimal. The internalisation of externalities has the effect of making the perceived local optimum coincide with the global optimum.

This policy implies that international regulation needs to be directed to the creation and maintenance of a *global premium* to investments in diverse resources. This is an additional return, created through funding by the international community, that will flow to owner-states investing in their diverse resources. There are only two logically distinct approaches to the creation of such premia: *international subsidy agreements* or *market regulation agreements*. However, there are a number of different forms (international parks, international management subsidies, intellectual property rights, producer cooperative agreements, consumer purchasing agreements and resource exchanges) that either of these approaches may take. The following two sections survey the nature of these alternative forms of intervention.

International subsidy agreements

Although it is underinvestment that generally drives extinction, it is the specific costliness of particular resource requirements that is the proximate cause of extinction. That is, extinction is caused most directly by the refusal to allocate scarce societal resources to the lands or management that diverse resources require. It is the refusal to purchase these life-sustaining factors for certain forms of biological resources that is the direct cause of species' decline.

These sources of decline can be remedied through a *system of strategic international payments*. This system of payments would necessarily be conditional upon the owner-state's application of them to the purchase of the required factors for specified diverse resources. That is, the payments would be restricted to use for purchases of land and management for a particular resource or region.

A crucial feature of any international scheme of payments based on conditionality (here, payment conditional on specific application) is its necessarily dynamic nature. That is, it is only possible to restructure the owner-state's decision making process if the payment is offered at the end of each period that the state takes the specified action. The payment is the *global premium,* supplementing the state's return, received for pursuing the global rather than the local optimum.

An example of such an international factor-subsidy scheme might be an *"international parks agreement."* Such a programme would in effect buy the use of the land for the diverse resources that exist there. The role of the international community would be to pay the rental price of the land each year that the national park remained unconverted. In order to manage the park, it would probably be necessary for the international community to provide a subsidy for management services as well. If the international community wishes to maintain diverse resources at the lowest levels of exploitation (e.g., nonconsumptive activities such as tourism and filming), then it will be necessary to provide almost complete subsidies for the foregone land development opportunities and management services required.

However, the maintenance of a stock of diverse resources will often be compatible with a wide range of forms of resource utilisation, other than those that are of the lowest intensity. In that case the international subsidies may be reduced in accordance with the extent to which development opportunities are allowed in the region. This is, in essence, the *development rights* approach to diverse resource conservation; to the extent that the international community wishes to reduce the development intensity away from the local optimum (of high intensity conversion and use), then it must be willing to provide a stream of *ex post* payments to compensate for the foregone development.

Another example of an international subsidy scheme would be a *resource franchise agreement.* This would be a three-way agreement (between owner-state, international community and franchisee) in which the international community would provide a stream of rental payments to the owner-state in return for its agreement to offer a piece of land for use only for restricted purposes specified within the franchise agreement. That is, the land would be designated for use, but only for limited uses amounting to much less than complete conversion (e.g., extractive industries such as rubber-tapping and plant and wildlife harvesting). The land would then be franchised to an entity that is allowed to use it only for purposes compatible with the franchise. If the state fails to enforce the franchise agreement, it forfeits its year's rental fee.

The benefit of a franchise agreement over an international parks scheme is that it provides a stream of benefits to fund the management of the operation. Under the franchise agreement, it is the responsibility of the franchisee to provide management services from the returns it generates in its operation of the franchise. In this fashion, the international community is able to "contract out" the provision of management services, while (through the rental fee and restricted use clauses in the franchise agreement) it retains the possibility of moving the local optimum closer to the global.

Therefore, a franchise agreement is simply a more generalised form of the international park agreement required for the acquisition of all development rights in a region. A franchise agreement could be used to specify the precise range of activities allowed and disallowed in the region, and it would be required of the international community to provide the "global premium" that is necessary to make up the difference between what an entity would bid for the franchise and what is the value of the land in unrestricted use.

Market regulation agreements

The alternative to the direct purchase of development rights is the subsidisation of diverse resource production, in order to alter the perceived benefits of conversion. That is, the owner-state considering conversion will balance the comparative benefits from the land in its various uses. One means by which the decision-making process might be biased towards the naturally occurring slate of resources is the enhancement of the returns from these resources.

Rent appropriation is a policy based on assuring that the owner-state receives the full value from its diverse resources. At present, this is the opposite of what is occurring with respect to a wide range of these resources. African countries were capturing about 5% of the value of their raw ivory exports at the height of the ivory trade (Barbier et al. 1990). Tropical bird harvesters around the world acquire between one and 5% of the wholesale value of the animals (Swanson 1992c). This is even true with regard to many exports of tropical forest products (Repetto and Gillis 1988). The owners of diverse resources under these circumstances are holding only the legal, and not the beneficial, rights of ownership. Combining the two sets of rights within the same entity will greatly enhance the perceived benefits from diverse resource management.

At present, international policies regarding diverse resource trade are the opposite of a rent appropriation system. International policy regarding poorly managed resources is to attempt to destroy the value of the resource (as is the case with various "trade bans" under the CITES convention) or to bring the resource into domestication (as with the relocation of the live stock for the tropical bird industry to the developed countries). Both approaches are geared equally to the destruction of the incentives for investments in diverse resource habitats.

An international policy regarding diverse resource trade must instead be based upon a constructive approach. That is, it must be used for the *maximisation of the rental value* of the resource combined with the *targeted return* of that value to those states investing in their diverse resources. This approach provides compensation for those states already investing in the management of their diverse resources, and it provides incentives to those states not so investing.

For example, the CITES installed ban on the ivory trade was a worst case for the African elephant. Although the continental populations of the elephant had declined by half, these populations had fallen precisely in those states that had not invested. In other states that had invested heavily in the species, such as Zimbabwe (with per kilometre investments 10 or 20 times as great as those in the noninvesting states), the elephant population had vastly increased over the same period (by 100% in Zimbabwe) (Swanson 1993a).

A blanket ban on the ivory trade provided precisely the wrong incentive structure for the African states. The imposition of a ban proved that the states engaging in elephant overexploitation were right; there was no future to be had from investments in this diverse resource.

The correct international incentive structure would do the opposite; it would provide *premia* to the states investing in their diverse resources and *penalties* to those who do not. This would be the role of a *sustainable wildlife trade exchange*. As with any exchange, it would be developed to discriminate between "good" (investing) and "bad" (noninvesting) suppliers and allow only the former to sell on the exchange; then, any consumer of the product would know that in purchasing from the exchange, the funding would flow to an owner-state that invests in the resource. In addition, to the extent that the consumer states agreed and enforced purchases solely from the exchange, the result would be greatly enhanced prices for the supplier states listed on the exchange. This *exchange-based price differential* would then constitute the premium for investing owner-states and the penalty for the noninvesting.

The crucial element to this approach is the creation of a price differential for investing owner-states through market regulation in the consumer-states. Restrictions on which suppliers consumers are allowed to make their purchases from will always create a premium for the favoured suppliers. A *global premium* for the sustainable producers of flows of diverse resources may be created through any scheme generally directed to this purpose.

Another example of such a scheme would be a *genetic resource (intellectual property) right regime*. This regime would also allot specific markets in consumer states to compensate for diverse resource investments, but the connection between the market and the investment would be less direct (as compared with stock investments that directly generate tangible flows).

The role of any form of intellectual property rights regime is to provide a basis for compensating investments in stocks that do not generate directly compensatable flows; specifically, intellectual property regimes generally reward inappropriable investments in information with rights in discrete markets. A concrete example is the innovation of the optimal sized racquet head, developed from a programme to determine the optimal trade-off between wind resistance (too large of a head) and required accuracy (too small of a head). The inventor of the "oversized" tennis racquet determined that a racquet of 117.5 cm^2 was optimal for tennis. In fact, this represented an investment in the creation of pure information that would not have been appropriable through marketing of tennis racquets (because other sellers would have immediately entered the market with the same head). Therefore, the intellectual property rights regime awarded this inventor with a protected market right in all racquet head sizes between 100 cm^2 (the original size of a tennis racquet head) and 135 cm^2. This protected market then acted as compensation for the investment in information creation made by this inventor.

It is equally possible to link protected markets to investments in diverse resources stocks, because these stocks also feed into various industries in an indirect and usually inappropriable fashion. For example, many pharmaceutical innovations are developed from a starting point of knowledge derived from the biological activities of natural organisms. When a new start is required, it is often initiated by returning to the uncharted areas of biological activity (unknown plants and insects), but after the long process of product development and introduction, there is no compensation for the role played by the diverse resource in initiating the process.

A genetic resource right system could be constructed that would be analogous to an intellectual property right system. That is, it would require a royalty payment to the owner-state investing in the maintenance of diverse resources that are made available for prospecting by various industrial concerns. This royalty would be based in a protected market right for return of some share of the revenues from the ultimately marketed products to the investors in the information at the foundation of the product (Swanson 1995b).

In summary, the idea of consumer market agreements is to allocate these markets only to those owner-states investing in their diverse resources. The owner-states that choose to mine their diverse resources will otherwise drive down the prices, and rents, available to all states providing diverse resource flows. An agreement to restrict consumer markets to those owner-states that invest in their diverse resources creates a price differential: a price premium target to all sustainable producers and a price penalty target to all nonsustainable. Such a mechanism might be used in a wide variety of circumstances, where the stock-related investments are directly linked to the final product (e.g., an ivory exchange) and where the stock-related investments are less directly linked to the final product (e.g., a genetic resource right regime).

8.7   Conclusions

The global problem of biological diversity losses may be viewed as the problem of halting the global conversion process at the globally optimal point in time (and space). Individual state decision-making cannot achieve this because there is no recognition of the countervailing benefits to the force for conversion. The marginal value of biological diversity contains a component of nonappropriable benefits that flow to the global community at large.

There is no doubt that there is a value to diversity. The decline of global stocks of diverse resources necessarily results in losses of global insurance and informational services. These are services that flow to the entire global community, and this is the fundamental reason for their importance and for their decline.

It is only international intervention that can halt this conversion process. However, the international community can act on this problem only through the instrument of the host state. The nature of international action must be the inducement of an otherwise nonexistent pattern of investments by the Southern states.

The global solution to the global biodiversity problem is investment in international institutions by the global community. These investments must establish mechanisms which provide incentives for host states to invest in diverse resources. The nature of these incentives will be the rechannelling of the global values of diverse resources back to the host states themselves. Such a system of intervention provides the necessary incentives to halt the global conversion process at the optimal point, while maintaining the capability of the marginal states to pursue development (albeit down diverse pathways).

This chapter has identified the general nature of two forms of intervention: "international subsidy agreements" and "market regulation agreements." A wide range of specific instruments fall under these two general categories: international parks, franchise agreements, producer cooperatives, wildlife trade regimes and genetic resource rights. The global community must invest in a range of these various institutions in order to halt the homogenising conversion process that threatens global biodiversity.

Therefore, the global diversity problem may be viewed as the externality emanating from the pursuit of the same, narrow pathway to development by successive states. As the final states embark on this same avenue, the commonality in the methods of production employed jeopardises the biological production system itself. The solution must then be the encouragement of the remaining uncommitted states toward diverse pathways of development. However, global institutions for development have always been constructed around the ubiquitous model, and they have ignored the alternatives. The international regulation of biodiversity decline requires the construction of international institutions focused upon alternative development routes. This chapter has identified and analysed a few of these alternatives.

CHAPTER 9

Policies to control tropical deforestation: trade intervention versus transfers

Edward B. Barbier and Michael Rauscher*

9.1   Introduction

Concerns about tropical deforestation have led to an increased focus on the role of the timber trade in promoting forest depletion and degradation. Recent reports suggest a marked increase in tropical deforestation in the 1980s, with the overall rate doubling from 0.6% in 1980 to 1.2% in 1990 (Dembner 1991). However, the deforestation rate varies across regions, with an estimated annual rate for Latin America of only 0.9% compared with 1.7% for Africa and 1.4% for Asia.

Despite concern over the state of tropical deforestation and its implications for global welfare, several recent studies have indicated that the tropical timber trade is not the major *direct* cause of the problem – perhaps less than 10% of total deforestation – rather, it is the conversion of forests for agriculture that is much more significant (Amelung and Diehl 1991; Barbier et al. 1994b; Binkley and Vincent 1991; Hyde et al. 1991). Nevertheless, it is clear that current levels of timber extraction in tropical forests – both open and closed – exceed the rate of reforestation (WRI 1992). Less than 1 million hectares, out of an estimated total global area of 828 million hectares of productive tropical forest in 1985, was under sustained yield management for timber production (Poore et al. 1989). Moreover, timber extraction has a major *indirect* role in promoting tropical deforesta-

* We would like to thank Joanne Burgess for contributing to discussions that resulted in this paper, which were held at the Beijer Institute of Ecological Economics of the Royal Swedish Academy of Sciences. We acknowledge the financial support given by the Beijer Institute and the hospitality of its staff, and in particular we are grateful to Prof. Karl-Göran Mäler for inviting us to use the facilities at Beijer and to Charles Perrings, Director of the Biodiversity Programme. Additional support for Edward Barbier's participation in the research was also provided by the International Tropical Timber Organization, under contract no. PCM(XI)/4, "Economic Linkages Between the International Trade in Tropical Timber and the Sustainable Management of Tropical Forests." We are grateful to comments provided by two anonymous referees; however, the usual disclaimer applies.

tion by opening up previously unexploited forest, which then allows other economic uses of the forests such as agricultural conversion to take place (Amelung and Diehl 1991; Barbier et al. 1991). For example, in many African producer countries, around half of the area that is initially logged is subsequently deforested, while there is little, if any, deforestation of previously unlogged forested land (Barbier et al. 1994b).

Some of the environmental values lost through timber exploitation and depletion, such as watershed protection, nontimber forest products, recreational values, etc., may affect only populations in the countries producing the timber. Concerned domestic policymakers in tropical forest countries should therefore determine whether the benefits of incorporating these environmental values into decisions affecting timber exploitation balance the costs of reduced timber production and trade, as well as the costs of implementing such policies. The socially "optimal" level of timber exploitation and trade is one where the additional domestic environmental costs of logging the forests are "internalised" in production decisions, where feasible. Designing policies to control excessive forest degradation is clearly complex and requires careful attention to harvesting incentives. As recent reviews suggest, many domestic policies do not even begin to approximate the appropriate incentives required to achieve a socially optimal level of timber harvesting. More often than not, pricing, investment and institutional policies for forestry actually work to *create* the conditions for short-term harvesting by private concessionaires, and in some instances, even *subsidise* private harvesting at inefficient levels.[1] Over the long term, incentive distortions that understate stumpage values and fail to reflect increasing scarcity as old growth forests are depleted can undermine the transition of the forestry sector from dependence on old growth to secondary forests and the coordination of processing capacity with timber stocks (Binkley and Vincent 1991).

Increasingly the world's tropical forests, including their remaining timber reserves, are also considered to provide important "global" values, such as a major "store" of carbon and as a depository of a large share of the world's biological diversity (Pearce 1990; Reid and Miller 1989). Similarly, even some "regional" environmental functions of tropical forests, such as protection of major watersheds, may have transboundary "spillover" effects into more than one country. But precisely because such transboundary and global environmental benefits accrue to individuals outside of the countries exploiting forests for timber, it is unlikely that such countries will have the incentive to incur the additional costs of in-

---

[1] For example, see Barbier et al. (1991), Gillis (1990), Hyde et al. (1991), Pearce (1990) and Repetto (1990).

corporating the more "global" environmental values in forest management decisions. Not surprisingly, sanctions and other interventions in the timber trade are one means by which other countries may seek to coerce timber producing countries into reducing forest exploitation and the subsequent loss of environmental values. In addition, trade measures are increasingly being explored as part of multilateral negotiations and agreements to control excessive forest depletion, to encourage "sustainable" timber management and to raise compensatory financing for timber producing countries that lose substantial revenues and incur additional costs in changing their forest policy.

However well-intentioned they may be, both domestic and international environmental regulations and policies that attempt to "correct" forest management decisions may have high economic, and even "second order" environmental, costs associated with them (Barbier et al. 1994b). There is increasing concern that the potential trade impacts of environmental policies that affect forestry and forest-based industries may increase inefficiencies and reduce international competitiveness. Moreover, the trade impacts of domestic environmental regulations may affect industries in other countries and lead to substantial distortions in the international timber trade. The overall effect on the profitability and efficiency of forest industries may be to encourage forest management practices that are far from "sustainable." Careful analysis of both domestic and international environmental policies affecting forest sector production and trade is therefore necessary to determine what the full economic and environmental effects of such policies might be.

The following model has been developed to facilitate analysis of the impact of trade interventions and international transfers on a timber-exporting tropical forest country. The main focus of the analysis is on how these impacts relate to the country's decisions to produce timber, or processed goods that are based on timber extraction, and thus the rate of tropical deforestation. Barbier and Rauscher (1994) extend the basic model to include the implications of market power and foreign asset (or debt) accumulation by the timber-exporting country. Here, we focus on the basic model, and support it with empirical evidence from a recent study of the linkages between the timber trade and tropical deforestation (Barbier et al. 1994b).

The basic model is similar to the one developed by Rauscher (1990), but it differs in two important respects. First, timber products can either be exported or consumed domestically, and the export earnings are used to import domestic consumption goods from abroad. This facilitates analysis of the *trade diversion* effect to domestic consumption of a policy intervention in the international timber market. Second, it is assumed that the tropical forest has positive *stock externalities* in the form of watershed protection, genetic diversity, microclimatic functions, etc. which directly affect the overall welfare of the country. The analytical results derived from the model are clearly affected by this assumption that the forest stock has some direct social value in addition to its use as a timber resource.

The model is also simplified in some important respects. Domestic capital accumulation and any tropical reforestation efforts are not modelled, and no other production or trade sectors are included, as this would complicate the analysis without providing deeper insights into the role of timber trade policy interventions in tropical deforestation. Initially, it is assumed that the country is a price taker in the international timber market and that trade is balanced. Barbier and Rauscher (1994) discover that relaxing these assumptions to allow for market power and foreign asset accumulation (or debt) does not affect significantly the main conclusions concerning the role of the timber trade in tropical deforestation and the relative effectiveness of trade policy interventions as opposed to international transfers in terms of reducing deforestation.

9.2   The model of a timber exporting tropical forest country

For the basic model, the following variables, parameters and functions are defined:

$N(t)$	tropical forest stock
$q(t)$	tropical timber logs extracted or commodities produced (log-equivalents)[2]
$x(t)$	tropical timber logs/products exported (log-equivalents)
$q(t) - x(t)$	domestic consumption of logs/products
$c(t)$	consumption of imported goods
$g(N(t))$	regeneration function of tropical forests
a	deforestation rate, per unit of (log-equivalent) timber extracted
p	terms of trade, p_x/p_c
δ	social rate of discount

Notation is simplified by omitting the argument of time-dependent variables, by representing a derivative of a function by a prime, by employing

---

[2] Processed timber products are assumed to be converted to log-equivalents.

numbered subscripts to indicate partial derivatives of a function, and by denoting the time derivative and growth rates of a variable by a dot and hat, respectively.

The tropical forest country is assumed to maximise the present value of future welfare, W

$$\underset{(q,x,c)}{\text{Max}} \ W = \int_0^{\infty} U(q - x, c, N) \, e^{-\delta t} dt \tag{9-1}$$

subject to

$$px = c \tag{9-2}$$

$$\dot{N} = g(N) - aq \tag{9-3}$$

$$N(0) = N_0 \quad \text{and} \quad \lim_{t \to \infty} N(t) \geq 0 \tag{9-4}$$

$$N^{\max} > N^{\min} > 0, \quad g(N^{\min}) = g(N^{\max}) = 0, \quad \text{and} \quad g''(N) < 0. \tag{9-5}$$

The control variables of the model are q, x and c. The additively separable utility function, U, is assumed to have the standard properties with respect to its partial derivatives, $U_i > 0$, $U_{ii} < 0$ ($i = 1,2,3$). Equation (9-2) is the initial trade balance assumption. Equations (9-3) to (9-5) are the standard renewable resource constraints, which suggest that any deforestation due to timber extraction net of regeneration will lead to a decline in the tropical forest stock.

9.3   Optimality conditions

The Hamiltonian of the optimal control problem above is

$$H = U(q - x, px, N) + \lambda[g(N) - aq] \tag{9-6}$$

where λ is the costate variable or the shadow price of the tropical forest. Assuming an interior solution, the maximum principle yields the following conditions

$$U_1 = \lambda a \tag{9-7}$$

$$U_1 = pU_2 \tag{9-8}$$

$$\dot{\lambda} = (\delta - g')\lambda - U_3 \tag{9-9}$$

where $U_1 = \partial U / \partial (q - x)$, $U_2 = \partial U / \partial c$ and $U_3 = \partial U / \partial N$. Equation (9-7) indicates that, along the optimal trajectory, the marginal value of extracting one unit of timber (in terms of domestic consumption), U_1, must equal its marginal depletion cost, λa. Since extraction costs are zero in the model, the latter costs are *user costs,* the future stream of timber income foregone from extracting a unit today. Equation (9-8) indicates that, if international terms of trade are given, the relative marginal value of domestic timber to imported good consumption must be equated with the terms of trade, *p.* Finally, Equation (9-9) yields a standard renewable resource dynamic condition for denoting the change in the value of the tropical forest stock when that stock also has direct value, as represented by U_3. As this condition is important for the analytical results of our model, we state its interpretation formally as

Proposition 1. *The rate of change in the shadow price of the tropical forest, λ equals the difference between the opportunity cost of holding on to a unit of the forest, $(\delta - g')\,\lambda$, and the marginal social value of that unit, U_3.*

Since the Hamiltonian is concave in (q,x,c,N), the preceding conditions are also sufficient for an optimum. By combining Equations (9-7) and (9-9) one obtains

$$\frac{\dot{q} - \dot{x}}{q - x} = 1/\eta_1(\delta - g' - aU_3/U_1) \tag{9-10}$$

where η_1 is the elasticity of marginal utility, U_1. Utilising Equation (9-10) and Conditions (9-3), (9-4), (9-5), (9-7) and (9-8), one can solve for an optimal saddle path and the long run equilibrium. As the system is in equilibrium when the user costs, the felling rate, domestic and imported consumption and the forest stock are constant, the equilibrium can be characterised by the following system of equations

$$U_1 - pU_2 = 0 \tag{9-11}$$

$$(\delta - g')U_1 - aU_3 = 0 \qquad \text{for } \dot{q} = \dot{x} = 0 \tag{9-12}$$

$$g(N^*) - aq^* = 0 \qquad \text{for } \dot{N} = 0 \tag{9-13}$$

where the equilibrium values of N and q are denoted by asterisks. Total differentiation of the system of Equations (9-11)–(9-13) with respect to (q,x,N) can lead to characterisation of the equilibrium state and trajectories leading to that state. It can be demonstrated that the determinant D of the Hessian matrix of above system must be negative in order for there to be a unique equilibrium in (q,N) space; i.e., the requirement for the curve $\dot{q} = 0$ to be positively sloped and to cut the curve $\dot{N} = 0$ from below is that $D < 0$ (see appendix). This implies that the equilibrium is a saddle point, and the saddle path is positively sloped. Using these results, the optimal solution is represented graphically in Figure 9.1. The solution suggests the following proposition.

Proposition 2. *The equilibrium forest stock, N*, is determined by equating the social discount rate, δ, and the rate of return from holding on to the forest stock, g′(N*) + aU₃/U₁.* $g'(N^*) + aU_3/U_1$. *The equilibrium will occur to the left of the maximum sustainable yield* (MSY) *if* $g'(N^*) > 0$, *and it will occur to the right of the* MSY *if* $g'(N^*) < 0$.

Thus, as depicted in Figure 9.1, if the initial level of the forest stock is high (i.e., $N_0 > N^*$), then the economy will deforest some of this stock through timber extraction.

Moreover, Equations (9-11)–(9-13) also imply

Proposition 2a. *If* U_3 *were* $= 0$, *then the equilibrium forest stock, N**, would occur at* $\delta = g' > 0$.

As shown in Figure 9.1, it must follow that $N^{**} < N^*$. That is, if the economy values only tropical timber then it will tolerate a lower level of tropical forest in the long run than if it also considers the other goods and services provided by the forest.

9.4   Comparative static analysis: trade interventions versus transfers

Comparative static analysis of the long run equilibrium can be employed to indicate what impacts reductions in the terms of trade for tropical timber and forest products, either through import bans, tariffs or other controls, may have on the tropical forest country's decision to deforest. As noted, the model already suggests that the tropical forest has positive *domestic* externalities in the form of watershed protection, genetic diversity, microclimatic functions, etc. which directly affect the overall welfare of the country. However, the *international* externalities, such as the role of the forests as a "store" of biodiversity and carbon and their "macro" climatic functions, are essentially ignored by domestic policymakers. Thus it can be assumed that intervention in the global timber market is motivated by the international community – notably tropical timber importers – attempting to force the country to "internalise" the global values ascribed to its tropical forest that are lost through the depletion arising from timber production.

A ban on tropical timber imports or the imposition of import taxes that discriminate against trade in tropical timber reduces the terms of trade. In the model, a reduction in the terms of trade, *p*, has the following impacts on the long run equilibrium forest stock of the timber exporting country

$$\frac{dN^*}{dP} = \frac{[1 + \eta_2]U_2\,[-a(\delta - g')U_{11}]}{D} \tag{9-14}$$

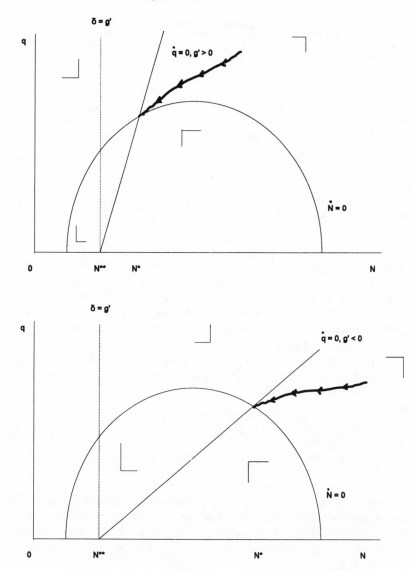

Figure 9.1. The long-run equilibrium of the tropical forest economy.

where η_2 is the elasticity of marginal utility, U_2, with respect to imported consumption goods. If the absolute value of η_2 is large, then marginal welfare in the economy is highly responsive to a change in imported consumption goods, c. We characterise this condition as "import dependency." The following proposition therefore results from Equation (9-14).

Table 9.1 *Price effects on equilibrium values*

	$|\eta 2|>1$	$|\eta 2|<1$
$g' > 0$	$dq^*/dp > 0$	$dq^*/dp < 0$
	dx^*/dp ?	dx^*/dp ?
$g' < 0$	$dq^*/dp < 0$	$dq^*/dp > 0$
	$dx^*/dp < 0$	$dx^*/dp > 0$

Source: See text.

Proposition 3. *A decrease in the terms of trade,* p, *will actually reduce the long-run equilibrium forest stock,* N*, *if the country is import dependent* ($|\eta_2| > 1$).

This would suggest that the use of timber trade interventions by importing countries may under certain economic conditions be counterproductive in their effects. Timber exporting countries may not always respond by reducing exploitation of their forests; rather, as indicated in our model import dependency and other economic considerations may lead to the opposite result in the long run.

The "second best" nature of trade interventions can be further seen through the effects of reducing the terms of trade on the long run level of timber extraction and exports

$$\frac{dq^*}{dp} = \frac{-(U_2 + pxU_{22})g'(\delta - g')U_{11}}{D} \tag{9-15}$$

$$\frac{dx^*}{dp} = \frac{[1 + \eta_2]U_2[a(g''U_1 + aU_{33}) - g'(\delta - g')U_{11}]}{D} \tag{9-16}$$

Changes in timber extraction and exports will clearly depend not only on the degree of import dependency as represented by η_2 but also on the relationship of the original equilibrium forest stock, N^*, with respect to the *MSY* (see Proposition 2 and Figure 9.1). The effects are summarised in Table 9.1.

If trade interventions by importing countries do not always achieve the desired effect of encouraging timber exporting nations to reduce exploitation of their tropical forest stock, then an alternative policy may be the provision of an "international transfer" of funds to encourage exporting countries to forgo the income earned from forest exploitation. Essentially, the rest of the world is "subsidising" tropical forest countries to conserve rather than cut down their trees. For example, the estimated international financing required could be in the region of US$ 0.3 to 1.5 billion annually (see Section 9.5).

A large transfer of international funds to tropical forest countries to assist with sustainable forest management and forest conservation has the effect of "freeing up" domestic financial resources for other purposes. In our model, an international transfer or subsidy can be represented by an increase in foreign exchange available for consumption of imported goods; i.e., it supplements timber export earnings. Thus (9-2) now becomes

$$px + s = c \qquad (9\text{-}2a)$$

where s is the amount of the international transfer, or subsidy. The comparative statics of Equation (9-11) in the system (9-11) to (9-13) becomes

$$U_{11}dq + [-U_{11} - p^2 U_{22}]dx = pU_{22}ds \qquad (9\text{-}17)$$

Thus an increase in international transfers, s, has the following impact on the long-run equilibrium forest stock, N^*

$$\frac{dN^*}{ds} = \frac{pU_{22}[-a(\delta - g')U_{11}]}{D} > 0 \qquad (9\text{-}18)$$

The following proposition therefore holds

Proposition 4. *A direct international transfer,* s, *will increase the long run equilibrium forest stock,* N*, *unambiguously.*

In comparing Propositions 3 and 4, it is clear that the comparative statics of the long run equilibrium clearly favour international transfers as the preferred method of inducing tropical timber exporting countries to conserve their tropical forests.

9.5   Empirical evidence: trade interventions versus transfers

These conclusions of our theoretical model are supported by empirical evidence from a recent study by the London Environmental Economics Centre (Barbier et al. 1993b; hereafter referred to as the "LEEC study"). This study examined both the implications of timber trade interventions, such as trade bans, quantitative restrictions and taxes, and international transfers for encouraging sustainable management of timber production in tropical forest countries. The empirical results for each policy intervention are discussed in turn.

Tropical timber trade bans

Pressures from environmental groups and consumer-led boycotts in developed market economies are leading to serious consideration of a *complete* ban on the import of all tropical timber products, or at least a *selective* ban on those products that are not "sustainably produced." However, despite their popular appeal, the use of such bans would not be appropriate for encouraging sustainable management in tropical timber exporting countries. There are several reasons for this.

First, producer countries argue, with some justification, that a ban on tropical timber products is *discriminatory;* i.e., similar rules do not apply to the temperate timber trade. It is unlikely that they would allow such a policy option to be sanctioned by any multilateral forum. As will be discussed, without the cooperation of producer countries, an import ban would most likely be *arbitrary* and *unworkable.*

To extend the ban to include both tropical and temperate product trade would be even more unfeasible. Those developed market economies that produce temperate timber also tend to be the most under pressure from environmental groups to instigate a tropical timber import ban. Given that the global temperate market is much larger and highly competitive, the governments in these countries would probably resist hurting the prospects of their own forest industries by extending the ban to cover all timber product trade.

More importantly, there is substantial evidence suggesting that an import ban on tropical timber products would be *ineffective* in reducing either tropical deforestation or the trade in "unsustainable" timber – and may be even *counterproductive.* Timber production is not the major cause of tropical deforestation. Not all (and a declining share) of the tropical timber produced is for export and an increasing share of tropical timber exports is being absorbed in South-South trade. This would suggest that, in response to a tropical timber ban imposed by current importers, major tropical timber exporters (e.g., in Southeast Asia) may be able to divert some timber supplies to domestic consumption, or to newly emerging export markets, fairly easily. For those tropical forest countries where timber exports are not significant, and are not a major factor in deforestation (e.g., Latin America), a ban may have little impact on timber management or overall deforestation.

In fact, to the extent that a tropical timber import ban does affect the export of tropical timber products substantially, it would have little impact on the economic *incentives* for sustainable management at the concession level, and may actually encourage poor management practices. Domestic and market policy failures in tropical forest countries affect the "internalisation" of the user and environmental costs of timber harvesting by concessionaires. Major policy changes in the forestry sector will be required to address these issues, yet by imposing trade bans importing countries may reduce their *political leverage* in influencing policy makers in producing nations.

Although in the short run a trade ban may reduce pressure on tropical forests through lower production for the trade, in the medium and long run a ban is likely to have a detrimental impact. By eliminating the gains from trade, a ban on tropical timber imports would decrease the value derived from timber production and thus actually reduce the incentives for tropical forest countries to maintain permanent production forests. Faced by declining export prospects and earnings from tropical timber products, developing countries may decide to convert more forests to alternative uses, notably agriculture. Thus the effect of the ban may be to reduce log production and exports, but may actually increase overall tropical deforestation in the medium and long term.

As part of a recent tropical timber trade study the effects of a total import ban on a major tropical timber producer, Indonesia, were simulated (Barbier et al. 1994a). According to the results of the analysis, an import ban would have a devastating impact on Indonesia's forest industry in the short term (see Table 9.2). Although there would be significant diversion of plywood and sawnwood exports to domestic consumption, this would be insufficient to compensate for the loss of exports. Net production in both processing industries would fall. Given its export orientation, the plywood industry would be particularly hurt – reducing its output by over 40%. Net production losses in the sawnwood industry would be closer to 10%. The overall effect is to lower domestic log demand in the short term by around 25–30%.

The policy scenario is of course assuming that the import ban is 100% effective. It is unlikely that all importers of Indonesia's tropical timber products – many of which are also newly industrialising or producer countries with processing capacities – would go along with a Western-imposed ban. In any case, one would expect that over the longer term there would be some diversion of Indonesian plywood and sawnwood exports to either new import markets or existing markets that prove to be less stringent in applying the ban (e.g., other developing or newly industrialising countries).[3]

The long-term implications of an import ban on tropical deforestation are also not encouraging. Even if the ban is 100% effective in the short term, any reduction in tropical deforestation resulting from lower levels of timber harvesting is likely to be short-lived. A total import ban would cause a major diversion of Indonesian timber products to meet domestic demand. Although in the short-term net production of wood products,

---

[3] These effects cannot be captured explicitly in the model.

Table 9.2 *Indonesia timber trade and tropical deforestation simulation model*

Key variables	Total import ban[a]	1% revenue raising import tax[b]	5% revenue raising import tax[c]
1. Prices (Rp/m³)			
Log border-equivalent price (unit value)	—	- 0.17	- 0.82
Sawnwood export price (unit value)	—	- 0.11	- 0.54
Plywood export price (unit value)	—	- 0.21	- 1.03
2. Quantities ('000 m³)			
Log production	- 28.33%	- 0.04	- 0.19
Log domestic consumption	- 27.37%	- 0.04	- 0.18
Sawnwood production	- 10.64%	- 0.03	- 0.14
Sawnwood exports	- 100.00%	- 0.23	- 1.12
Sawnwood domestic consumption	30.01%	0.06	0.30
Plywood production	- 43.84%	- 0.04	- 0.22
Plywood exports	- 100.00%	- 0.10	- 0.51
Plywood domestic consumption	214.51%	0.23	1.12
3. Deforestation (km²)			
Total forest area	-	0.00[d]	0.01
Annual rate of deforestation	-	- 0.41	- 0.72

Note: Policy scenario: import ban and revenue raising taxes, percentage change over base case.
[a]Large price changes were used deliberately to constrain sawnwood and plywood exports to zero in this simulation and therefore are no longer endogenously generated by the model. Also, the functional form of the deforestation equation and its estimation using regional panel data imply that the large changes in log production associated with the import ban scenario cannot be used to predict reliably the effects on forest cover and deforestation. Thus, both price and deforestation effects are eliminated from this policy scenario simulation.
[b]A total of US$23.1 million (1980 prices) in revenue would be raised, with US$5.8 million and US$17.3 million from Indonesian sawnwood and plywood exports, respectively.
[c]A total of US$113.9 million (1980 prices) in revenue would be raised, with US$28.5 million and US$85.4 million from Indonesian sawnwood and plywood exports, respectively.
[d]A negligible increase over the base case forest cover of 53 sq km.
Source: Barbier et al. [1994a].

and thus log demand, would fall, this situation would not necessarily be sustained over the long run. Even if this is not the case, the ban may be ineffective in permanently reducing tropical deforestation because: (1) timber production is not the main source of deforestation in Indonesia; and (2) as the value of holding on to the forest for timber production decreases the incentives to convert more of the resource to alternative uses such as agriculture will increase.[4]

Many of the problems associated with a *complete* import ban on tropi-

[4] For further discussion, see Barbier et al. (1994a), Constantino (1990) and Vincent (1990).

cal timber products would also apply to a *selective* import ban on "unsustainably" produced timber alone. A selective import ban would also fail
to provide the appropriate economic incentives for sustainable management at the forest level, and may also be counter-productive, for the similar reasons outlined above for a complete ban, namely:

- *Diversion of trade to other markets* (domestic, export markets
  without bans, etc.). If these markets are for lower value products,
  producer countries may need to supply higher volumes of tropical
  timber to generate substantial earnings, thus leading to more
  pressure on timber resources.
- *Lower political leverage of importing countries* to influence forestry
  policies in producer countries.
- *Little positive reinforcement of the incentives for sustainable management.* Selective bans would have an immediate impact on a
  country's ability to derive value from timber production, and
  would act as a disincentive in the medium and long term to maintaining tropical forests for timber production, as opposed to conversion to agriculture and other uses.
- *Generate incentives to circumvent the ban.* The case study found
  that there is generally high elasticity of substitution for tropical
  timber products from different sources of origin, particularly for
  higher valued products such as plywood. Thus producer countries
  would gain significantly if they could "pass off" their "unsustainably" produced timber as "sustainably" produced.

Moreover, selective bans have the additional complication of the need for
a *non-arbitrary* and *workable* international certification process to distinguish "sustainably" versus "unsustainably" produced tropical timber
products. Given the lack of data on forest inventories and offtake levels,
such certification would be extremely difficult to establish in the near future without the cooperation of producer countries. However, there is little
incentive for producer countries to participate in this process if it leads to
an import ban on their products. Without this data, a selective ban could
not possibly succeed. In addition, an effective monitoring and evaluation
system would be required to enforce a selective import ban. No such
mechanism currently exists or is likely to be implemented in the next few
years. Again, cooperation by producer countries, which would be fundamental to the success of the verification system, is unlikely to be forthcoming. Finally, a selective trade ban is unlikely to be acceptable in the international political arena – particularly as producer countries will dismiss it
as *discriminatory.*

Quantitative restrictions

To some extent quantitative restrictions are similar to trade bans, but usually take a less severe form. However, a 100% restriction on the quantity of trade is effectively the same as a complete ban on all international trade in tropical timber. Thus, to a large extent, many of the problems associated with tropical timber trade bans also apply to quantitative restrictions.

Two quantitative restrictions are worth briefly examining: (1) quotas targeted to specific product categories, such as timber products derived from endangered species; and (2) quotas conditional on the level of sustainable offtake of specific species.

At the 8th Conference of Parties to the Convention on International Trade in Endangered Species (CITES), several tropical timber species were proposed for the Appendix I and II list of CITES (ITTC 1992b). Species given an Appendix I listing are effectively banned from commercial trade, or only authorised "in exceptional circumstances." Species listed on Appendix II are subject to strict regulation "in order to avoid utilisation incompatible with their survival." Of the species proposed for listing at the meeting, *Intsia spp* (merabu), *Gonystylus bancanus* (ramin), *Swietenia spp* (mahogany) and *Pericopsis elata* (*aformosia*) are internationally traded in significant volumes, including by producer countries that are members of ITTO. Not surprisingly, these proposals – which were put forward mainly by representatives of developed market economies at the Conference – were strongly opposed by the producer countries affected.

Studies of previous attempts by CITES to use quantitative restrictions and bans to regulate the trade in endangered species suggest that this approach has not been an effective means of control to date (Barbier et al. 1990; Burgess 1992). Problems of monitoring and enforcement are exacerbated when the producer countries affected either oppose this approach or do not receive adequate incentive (i.e., compensation) to participate. For example, at the 8th CITES Conference, Ghana and Cameroon opposed the proposed Appendix II listing of *aformosia,* and subsequently informed the Conference that the timber trade was outside the scope of their countries' CITES managing authorities (ITTC 1992b). The lack of compliance by producer countries undermines the effectiveness of such policy options.

Similar problems of incentives for management, enforcement and monitoring will exist for any import quotas based on the level of "sustainable" offtake of specific species that are imposed without the cooperation of tropical timber producing countries. For such a policy to be workable from the outset, cooperation from producer countries in determining the scientific and trade data necessary for establishing quotas for sustainable offtake and exports will be essential.

It might be possible to establish a more comprehensive trade mechanism that establishes sustainable offtake export quotas for those species that are endangered, thus offering an incentive to both consumers and producers to accept a controlled legal trade and to enforce it. A properly constructed trade mechanism for each endangered species, using economic instruments such as taxes and subsidies to manage trade where appropriate, could be designed to enable sufficient profits from the trade to be channelled back into producer states to encourage management efforts, and to support improved monitoring of harvesting and export activities (Burgess 1992). However, full acceptance of and adherence to any such mechanism by affected producer countries are essential to its success in terms of promoting compliance with export regulations and on cooperation between exporters and regulating agencies.

In sum, the use of quantitative restrictions to regulate the trade in tropical timber products can suffer many of the same problems associated with complete and selective trade bans. Although banning or controlling the trade in specific timber species may be a necessary short term response if the species are endangered, more effective and innovative long term solution for management of the trade are required. A more comprehensive trade mechanism that establishes sustainable offtake export quotas for endangered tropical timber species would offer a better incentive to both importers and exporters to accept a controlled legal trade and to enforce it.[5]

Trade taxes

An international trade tax would be a powerful means of increasing the costs associated with trading tropical timber products. Such a tax could be implemented at either the exporting or the importing end of the international tropical timber trade. In order for this tax to be effective in achieving a reduction in the amount of tropical timber products traded, it would have to be sufficiently high to stimulate producers and consumers to change their patterns of resource use. However, where the tax is implemented will affect the extent to which the costs of the tax are passed on to the producers or the consumers.

An international trade tax on tropical timber products suffers from the same sort of problems of implementation and effectiveness as faced by restrictions.

[5] As an encouraging sign, at the 8th Conference, the ITTO representative noted and welcomed the call made by several (government) parties for increased cooperation between CITES and ITTO (ITTC 1992b).

First, it is unlikely that tropical timber exporting countries would endorse an international tax that would discriminate against tropical timber products. As noted, these countries have employed their own export taxes on selective products (e.g., log exports, and more recently sawnwood exports). However, the main purpose of such policies has not been to *reduce the overall trade* in tropical timber products, or for that matter timber-related tropical deforestation, but to *change the pattern* of their tropical timber exports to higher valued products – thus stimulating wood-based processing industries.

Second, a tropical timber trade tax would in any case probably be *ineffective* in reducing timber-related tropical deforestation – for the same reasons outlined in the discussion of trade bans. In the short run, producer countries may increase their volumes of exports to maintain their level of foreign exchange earnings. Most likely, an overall trade tax might *reduce the incentives* for sustainable timber management, as the tax would lower the net returns from production and trade. As a result, the tax may actually prove to be *counterproductive* by encouraging the conversion of permanent production forests and any remaining "virgin" forest areas to alternative uses in the long run.

It is often argued that some of these problems associated with an indiscriminately applied tax on all tropical timber products could be avoided by a tax imposed selectively on "unsustainably" produced tropical timber products. For example, selective import duties on tropical timber products could be applied, but sustainably produced timber could be allowed in importing markets duty free. However, as argued in the case of selective bans and quantitative restrictions, there are obvious problems of certification, monitoring and enforcement that need to be sorted out – along with the political feasibility of such an approach. The issue of certification is an important one, and will be discussed in more detail.

The economic impact of an import tax on tropical timber was analysed in a policy simulation (Perez-Garcia and Lippke 1993). The policy scenario simulates the tax by assuming a 10% increase in the "transfer cost" of products reaching destination countries. The results suggest that:

- Some product exports are expected to be driven out by competition with domestic supplies.
- Log prices in consumer countries rise but fall in producer countries. As a consequence, most consumer countries reduce their imports of tropical timber logs.
- With such a large impact on log imports for their processing industries, consuming country demand for processed products results in a small increase in tropical plywood exports, at least ini-

tially. However, as this import demand slackens and as log scarcity in producer countries becomes binding, tropical plywood exports fall significantly after 1995.

- All tropical hardwood suppliers suffer a decline in production, at least initially. Indonesian production declines up to 2 million cm^3 compared to base case projections. However, the declines are not permanent in all regions. For example, Malaysia East (includes Sabah and Sarawak) experiences an initial decline of 4 mn m^3, but its log production generally exceeds the base case scenario after 1995.
- The net effect of log production declines in producing countries, distortions in trade patterns and falling producer prices for tropical timber harvesting will be less motivation to manage production forests sustainably, as well reduced revenue to finance these investments.

The timber trade model of Indonesia was also employed to see the effects of import taxes on a major exporter (see Table 9.2). The results suggest that tax rates beyond 1–5% would begin introducing major distortionary effects. However, such low levels of tax would clearly have little impacts on reducing deforestation in Indonesia. Their only purpose would be to raise revenue that could be then transferred to Indonesia and other tropical forest countries for forest conservation and sustainable management initiatives.[6] In addition, it would probably be more effective if Indonesia raised the money itself for "sustainable management" programmes by imposing revenue-raising export surcharges. As discussed below, the main issues are whether the amount of funds raised would be adequate and whether, in any case, they should be raised from sources outside of the trade altogether.

In sum, a trade tax on tropical timber products would not be an effective means of encouraging sustainable production forest management, and may even be a disincentive for sustainable management. A smaller surcharge imposed by exporting countries on their own trade may be more appropriate, but may not raise sufficient funds. In any case, as will be discussed next, the rationale for raising more substantial funds for international transfers may be more compelling – although politically difficult.

International transfers

The main rationale for providing financing for assisting tropical forest countries in moving towards sustainable forest management is that there

---

[6] See also Buongiorno and Manuring (1992).

is an important principle of *international compensation* at stake. There are essentially three reasons for this argument:

- It is often claimed that timber exporting countries receive an insufficient share of the returns from tropical timber product exports – at least to incur the additional harvesting costs and other economic impacts of sustainable timber management.
- Implementation of the forestry policies and regulations required to ensure the proper enforcement and monitoring of sustainable management of production forests will impose substantial additional costs on producer countries that they will find difficult to afford.
- To the extent that all nations benefit from the global external benefits resulting from sustainable management of large tracts of tropical forest lands, then the international community should compensate producing nations for the loss of potential income that they would incur by reducing tropical deforestation, timber sales and conversion of forest land to other uses.

As noted by Barbier et al. (1994b), the first point is difficult to substantiate. The issue may be less to do with whether the unequal distribution of revenues along the chain of trade *reduces the incentives* for sustainable management of the forest level but whether any excess revenues along the chain can be tapped for *additional funds* to assist sustainable management of tropical production forests. The second and third points are much more relevant to the argument for international compensation. It is now generally accepted, as well as enshrined in the Forest Principles of the 1992 UNCED Conference, that *compensating* tropical forest countries for their role in maintaining a resource that has value on a *global level* is a fundamental basis for multilateral policy action. However, this would suggest that compensation is needed by tropical timber producing countries for the income they may forego in protecting their forests and for the additional costs incurred in implementing sustainable management practices for their production forests. This has to be demonstrated empirically.

A policy simulation was conducted to indicate the additional economic impacts on tropical forest countries of "setting aside" some of their forest resource base (Perez-Garcia and Lippke 1993). Essentially, this was simulated by a reduced timber supply scenario where the inventory of commercial tropical hardwood resources is reduced by 10% – which is equivalent to land being taken out of product forests and permanently protected. The result is that severe shortages in log production and higher sawlog prices are experienced in tropical forest regions, notably in Malaysia and Indonesia. The model indicates that such reductions in supply would result in a loss of wealth for tropical timber producing countries. Over the long run, permanent set-asides would mean that the remaining production forest inventory could not support as high a level of sustainable harvest as under base case projections.

There are also indications that the additional costs required to implement sustainable forestry management policies and regulations are significant. Drawing on the work by Poore et al. (1989), a rough assessment of the resources needed by producer countries to attain sustainability by the year 2000 was conducted on behalf of ITTO (Ferguson and Muñoz-Reyes Navarro 1992). Although preliminary and very approximate, they show that an additional US$ 330.1 mn is required each year in order to assist producer countries of ITTO in attaining the Year 2000 Target. Moreover, at best the estimates are an indication of only the minimum financing required.

Sizeable though this figure may seem, it is less than the estimated amount required for sustainable management of *all* tropical forest resources. For example, Agenda 21 of the UN Conference on Environment and Development (UNCED) has estimated that international financing of over US$ 1.5 bn annually will be required by tropical forest countries to reduce deforestation (ITTC 1992b).[7] Using these two estimates as the broad "bounds" on the type of financing required, the LEEC study suggests that additional funds required by producer countries to implement sustainable management of their tropical forest resource to be in the range of *US$ 0.3 to 1.5 bn annually.*

As argued by Barbier et al. (1994b), there is also a strong rationale for additional funds to be made available to producer countries for sustainable forest management from sources outside of the tropical timber trade. Comprehensive international agreements, targeted financial aid flows and compensation mechanisms to deal with the overall problem of tropical deforestation may ultimately eliminate the need to consider intervention in the timber trade. Given that commercial logging is not the primary cause of tropical deforestation, such approaches would avoid unnecessary, and possibly inappropriate, discrimination against the timber trade. On the other hand, a comprehensive tropical forest agreement seems much

---

[7] The proposed international financing is for, specifically, sustaining the multiple roles and functions of all types of forests, forest lands and woodlands (US$ 860 mn p.a.) enhancement of the protection, sustainable management and conservation of all forests, the greening of degraded areas through forest rehabilitation, afforestation, reforestation and other rehabilitation measures (US$ 460 mn p.a.); and promoting efficient utilisation and assessment to recover the full valuation of the goods and services provided by forests, forest lands and woodlands (US$ 230 mn p.a.).

more difficult to negotiate and raises its own problems of workability and effectiveness.[8]

It is unlikely in the current global economic climate that a concerted international effort to increase substantially bilateral or multilateral aid flows for sustainable production forest management would be forthcoming, given the rapidly diminishing aid budgets of the developed economies. However, optimists who point to the mounting public concern for tropical forests in advanced industrialised countries would argue otherwise. Nevertheless, there still remains the possibility of designing new sources of financial assistance that are separate from existing developing country aid budgets. The Forestry Principles are effectively a stepping stone in that direction, and international commitments through the Tropical Forestry Action Plan (TFAP) and Global Environmental Facility (GEF) continue to reinforce the global interest in forestry and biodiversity protection. The case could be made for more comprehensive international agreements to raise revenues for sustainable management of tropical forests, including production forests. The main focal points of such agreements should be the development of alternative revenue-raising mechanisms other than trade interventions or reliance on existing aid budgets.

For example, Amelung (1991) argues the case for the establishment of an international rain forest fund, as proposed by UNEP, in order to avoid free-riding among nontropical countries. Similar arguments are put forward by Sedjo, Bowes and Wiseman, although their preference is for the establishment of a global system of marketable forest protection and management obligations (FPMOs). Over the long term, international negotiations leading towards an agreement for new revenue-raising mechanisms for sustainable management of tropical forests, including production forests, could be the most effective and equitable means for reducing global tropical deforestation. Although such negotiations are difficult and arduous, they may be the best hope of ensuring a global commitment to control tropical deforestation, as well as the sustainable management of production forests.

To summarise, the available empirical evidence would support the theoretical conclusions of our model: policy interventions in the form of trade bans, quantitative restrictions and taxes are not only economically inefficient but may also be counterproductive in achieving the objective of reducing tropical deforestation. Although international transfers to assist conservation and forest management efforts in tropical forest countries

---

[8] See Barrett (1990) for a general discussion of the difficulties involved in securing international environmental agreements.

are preferred, there are still major political and international obstacles to establishing such policies.

9.6   Conclusion

This chapter has examined several aspects of the links between the trade in tropical timber and deforestation from the perspective of an exporting country. The various versions of the model developed here have highlighted a number of important features of this linkage.

First, if the producer country values its tropical forests solely as a source of timber export earnings then it will aim for a smaller forest stock in the long run than if it also considers the other values provided by the forest. Understanding the full range of benefits accruing from their tropical forests, e.g., watershed protection, biodiversity, tourism, microclimatic functions, etc. is important to determining the direct social value of forest conservation.

Second, if importing nations want the exporting countries to conserve more of their forests, trade interventions appear to be a second-best way of achieving this result. Under certain conditions, they may be counterproductive. In contrast, international transfers, which in our model simply reduce the dependency of the producer country on the exploitation of the forest for export earnings, are more effective in promoting conservation of the forest stock.

Recent empirical studies have supported these analytical results of our model (Barbier et al. 1994b; Binkley and Vincent 1991; Hyde et al. 1991). Generally, these studies have also concluded that trade intervention is a "second best" option for controlling tropical deforestation. Nonetheless, the use of bans, tariffs and other trade measures to discourage "unsustainable" tropical timber exploitation continues to be advocated. As our chapter has attempted to show, sometimes the more simple solutions lead neither to straightforward, nor to the desired, results.

Appendix

Comparative static solution of the simultaneous equation system (11), (12) and (13) can be represented by

$$
\begin{pmatrix}
U_{11} & -U_{11} - p^2 U_{22} & 0 \\
(\delta - g')U_{11} & -(\delta - g')U_{11} - g''U_1 - aU_{33} & \\
-a & 0 & g'
\end{pmatrix}
\begin{pmatrix}
dx \\
dx \\
dN
\end{pmatrix}
=
\begin{pmatrix}
U_2 + pxU_{22} & 0 \\
0 & U_3 \\
0 & q
\end{pmatrix}
\begin{pmatrix}
dp \\
da
\end{pmatrix}
$$

Solution can be derived through the application of Cramer's rule. However, the key issue is the sign of the determinant of the Hessian matrix of the above system. We therefore derive this result. Having signed the determinant, the comparative static results are fairly straightforward.

The determinant D of the Hessian matrix is

$$D = g'p^2 U_{11} U_{22}(\delta - g') - a(U_{11} + p^2 U_{22})(g'' U_1 + a U_{33})$$

The right hand side is negative but the lefthand side is ambiguous since g' is unsigned. However, it can be shown that $D < 0$ is a requirement for the curve q to be positively sloped and to cut the curve N from below, which is a necessary (but not sufficient) condition for there to be a unique equilibrium in (q, N) space.

The implicit function rule applied to the above simultaneous Equation System (9-11), (9-12) and (9-13) yields

$$U_{11} dq = (U_{11} + p^2 U_{22}) dx$$

$$(\delta - g') U_{11} dq - (\delta - g') U_{11} dx = (g'' U_1 + a U_{33}) dN$$

$$- a\, dq = - g'\, dN$$

Substituting for dx yields

$$\left. \frac{dq}{dN} \right|_{\dot{q}=0} = \frac{(g'' U_1 + a U_{33})(U_{11} + p^2 U_{22})}{(\delta - g')p^2 U_{22} U_{11}} \quad > 0$$

$$\left. \frac{dq}{dN} \right|_{\dot{N}=0} = \frac{g'}{a}$$

It follows that

$$\left. \frac{dq}{dN} \right|_{\dot{N}=0} > \left. \frac{dq}{dN} \right|_{\dot{q}=0}$$

also implies that

$$a(U_{11} + p^2 U_{22})(g'' U_1 + a U_{33}) > g'p^2 U_{11} U_{22}(\delta - g')$$

and thus

$$D < 0$$

CHAPTER 10

On biodiversity conservation

Scott Barrett

10.1 Introduction

In an early round of the negotiations on the Convention on Biological Diversity, a number of fundamental principles were seen to guide achievement of the objectives of the Convention. The first two of these were, "The conservation of biological diversity is a [matter of] common concern of all humankind and requires cooperation by Contracting Parties," and "The Contracting Parties have as States the sovereign right to exploit their own biological resources pursuant to their own environmental policies. . . ."[1] Though these principles were subsequently edited,[2] they convey the essential message that biological diversity is a global public good, that all countries can potentially be made better off by cooperating, but that cooperation will only succeed if individual countries are made better off.[3] These principles also beg the questions: How should countries exploit their own biological resources? and How can cooperation be sustained by the Convention?

Importantly, the obligations of the final Convention apply unequally between "developed" and "developing" countries. Paragraph 4 of Article 20 reads:

> The extent to which developing country Parties will effectively implement their commitments under the Convention will depend on the effective implementation by developed country Parties of their commitments under the Convention related to financial resources and transfer of technol-

---

[1] Article 3, Paragraphs 1 and 2 of the Fifth Revised Draft Convention on Biological Diversity.

[2] The final Convention contains the first principle in its preamble, but with the clause "and requires cooperation by all Contracting Parties" omitted. The second principle is restated in Article 3 of the Convention, but with the word "biological" removed.

[3] This is a general requirement of international environmental agreements; see Barrett (1990).

ogy and will take fully into account the fact that economic and social development and eradication of poverty are the first and overriding priorities of the developing country Parties.

It is clear from this passage that the question of how a country's own biological resources should be exploited relates mainly to the developing countries, which must incorporate the conservation of biological diversity into their development planning. The question of how cooperation can be sustained by the Convention relates mainly to the developed countries, which must provide the financial resources needed to compensate developing countries for the (incremental) costs of conserving biodiversity. This paper explores these questions.

My focus will be narrow. I shall only consider the public goods aspect of biodiversity. This includes what the Brundtland Report (World Commission on Environment and Development 1987: 13) refers to as the ". . . moral, ethical, cultural, aesthetic, and purely scientific reasons for conserving wild beings." It also includes the value of the information embodied in biodiversity – including the blueprint of a species, which may be useful for developing synthetic chemicals and compounds or other goods, and the genetic material of a species, which may be useful in breeding programmes or in the new biotechnology.[4] The full value of these attributes of biodiversity cannot be captured by individuals or firms, or even by national governments. As Sedjo (1992, p. 27) notes, ". . . international law does not recognise property rights to wild species or wild genetic resource genotypes, and hence any rents associated with valuable natural genetic resources typically cannot be captured simply through domestic management of the resource, even by a national authority." Hence, biodiversity has both domestic and global public good characteristics.

Section 10.2 of this paper considers only the domestic conservation problem, and furthermore assumes that biodiversity conservation enters negatively in the production function – biodiversity conservation entails a cost measured as the social value of foregone production (consumption). This is an extreme assumption, but if it can be shown that biodiversity should be conserved under this assumption, then it can certainly be shown that biodiversity should be conserved in a setting which does recognise the other values of biodiversity.

The analysis is of a country in isolation, and can serve as a benchmark for Sections 10.3 and 10.4, which consider the global public goods aspect

---

[4] As an example of the former value, scientists have discovered that the fur of polar bears converts sunlight into heat with greater efficiency than even the most advanced solar technology. This discovery could redirect solar energy research, and reduce the cost of solar heating (Grow 1987). The latter value includes the use of germplasm from wild species to maintain the vitality of food crops.

of biodiversity conservation. In these sections, conservation does not enter only negatively in the production function. However, the analysis does recognise that global biodiversity conservation is not costless, and that there exist incentives for countries to free-ride on the conservation actions of others. The Brundtland Report made the point (1987: 160) that: "At the heart of the issue [of biodiversity conservation] lies the fact that there is often a conflict between the short-term economic interest of the individual nations and the long-term interest of sustainable development and potential economic gains of the world community at large." Section 10.3 models this conflict, while Section 10.4 seeks to determine whether the free-rider incentives facing countries can be overcome by careful design of an international agreement. Section 10.5 briefly summarises the main results of the paper.

10.2 Development and biodiversity conservation[5]

Consider an individual country. Its borders are fixed, and its total land area is normalised to equal one unit. At time t there exists D_t units of developed land and S_t units of wild land, with $D_t + S_t = 1$ (D_0 and S_0 are given and both are positive). Since the quantity of biological diversity can be taken to be an increasing function of wild land area, S_t might as well be interpreted as the stock of biological diversity.[6] Let L_t denote the population/labour force at time t. Further, let $L_0 = 1$, assume that population grows at rate ρ, and let $s_t = S_t/L_t$. Output is given by the production function $F(D_t, L_t)$. Assuming constant returns to scale, we can write $F(D_t, L_t) = L_t f(e^{-\rho t} - s_t)$ with $f' > 0, f'' < 0$. If there is Harrod-neutral technical progress and all output is consumed, per capita consumption will be given by

$$c_t = f(e^{-\rho t} - s_t)e^{\omega t} \tag{10-1}$$

[5] The model developed in this section is taken from Barrett (1992), but differs in two respects. The model presented here is more general insofar as it allows for exogenous population growth. However, this comes at the cost of assuming that the conversion of natural capital (wild land) into developed capital (what may be taken to be agricultural land) does not itself yield any income (of course, developed capital does yield an income). This last assumption greatly simplifies the analysis and is of no great consequence provided agricultural development is seen to be the main cause of biodiversity loss.

[6] Barrett (1994a) presents a model of optimal growth in which "biological diversity" is defined as "species diversity," and where the species-area relation from the biogeography literature is employed to relate wild land area to species. The constituents of S might also include support of ecosystem services and genetic distance. For analyses of how "diversity" might be defined and measured, see Weitzman (1992) and Solow, Polasky and Broadus (1993).

Let R_t be the rate at which wild lands are developed. Then $\dot{S}_t = -R_t$. Letting $r_t = R_t/L_t$ yields

$$\dot{s}_t = -r_t - \rho s_t \qquad (10\text{-}2)$$

Irreversibility means

$$r_t \geq 0 \qquad (10\text{-}3)$$

The objective functional is defined by the integral

$$\int_0^\infty U(c_t, s_t)\, e^{-(\delta - \rho)t}\, dt \qquad (10\text{-}4)$$

with $U_c > 0$, $U_{cc} \leq 0$, $U_s > 0$, $U_{ss} \leq 0$, and $U_{cs} = U_{sc} = 0$. Notice that S appears negatively in the production function (since $D = 1 - S$). There is a cost to conserving biodiversity in this model, for conservation implies some sacrifice in output. The only benefit is the intrinsic social worth of biodiversity (S appears positively in the instantaneous well-being function, U). Notice, too, that we must assume $\delta > \rho$ to ensure the existence of an optimal programme. Now, δ is the utility rate of discount, and one's ethics might demand that it equal zero, despite a rising population. However, Koopmans (1965) long ago demonstrated the illogic of this reasoning (see also Dasgupta and Mäler 1993).

Rather than derive the entire optimal development/conservation programme, here I wish to confine my attention to one particular aspect of this programme, namely the conditions under which the irreversibility constraint Equation (10-3) will bite. The question I wish to explore is this: under what conditions will it be optimal to conserve a greater quantity of S than is indicated by current costs and benefits? This qualitative question has been a central concern of the economics of environmental preservation (see, e.g., Krutilla and Fisher 1985).

To progress using a minimum of mathematics, it will prove helpful to begin by solving the problem assuming $\rho = \omega = 0$. This solution will serve as a benchmark for the analysis of irreversibility. Assuming an interior solution, it is easy to show that the economy should move instantly to the steady state

$$U_s = U_c f' \qquad (10\text{-}5)$$

Equation (10-5) says that wild lands should be developed until the marginal benefit of conserving biological diversity equals the associated marginal cost. Since population is constant and equal to one, $s = S$. Let S^+ denote the optimal steady state for this autonomous control problem (obviously, $S^+ \geq S_0$, $S^+ = s^+$).[7] Since U_s is decreasing in s and $U_c f'$ is increasing in s, S^+ is unique. One moves instantly to the steady state because foregoing some consumption today yields no additional consumption in the future and there is a cost to postponing consumption ($\delta > 0$).

Now consider the more interesting case where $\rho, \omega > 0$. Let S^* denote the optimal steady state for this non-autonomous control problem. Intuition suggests that if the LHS of Equation (10-5) rises relative to the RHS when $r_t = 0$, then it will not only be optimal to conserve some biological diversity, but to conserve a quantity $S^* > S^+$ ($S^* = S^+$ if $S^+ = S_0$), and to move instantly to S^* at time zero. This intuition turns out to be correct.[8]

To determine the condition which ensures $S^* > S^+$, it will prove helpful to specialise a bit further. Let $-U_{ss}s/U_s = \sigma \geq 0$, $-U_{cc}c/U_c = \eta \geq 0$ and $f = (e^{-\rho t} - s_t)^\alpha$ with $1 > \alpha > 0$. With $r_t = 0$, s_t falls over time at constant rate ρ. The marginal benefit of biological diversity conservation therefore rises at rate $\sigma\rho$. Upon differentiation and substitution, it can be shown that the marginal cost of conserving biological diversity changes at rate $\eta(\alpha\rho - \omega) + \omega - \rho(\alpha - 1)$. Hence, for $r_t = 0$, the marginal benefit of conservation rises relative to the marginal cost iff

$$\sigma\rho > -\eta(\omega - \alpha\rho) + \rho(1 - \alpha) + \omega \qquad (10\text{-}6)$$

If condition Equation (10-6) is satisfied, more biological diversity should be conserved than is indicated by current costs and benefits. If the inequality in Equation (10-6) is reversed, all biological diversity should be exhausted in the limit. This qualitative result depends on both factual parameters (ω, α and ρ) and ethical parameters (η and σ).[9]

In interpreting Equation (10-6), consider first the case where there is technical progress ($\omega > 0$) and no population growth ($\rho = 0$). Then condition Equation (10-6) reduces to $\omega (\eta - 1) > 0$. Now, one might think that technical progress would work against conservation, since the additional consumption obtained by conversion would be increasing over time. However, consumption derived from the stock of developed capital will also be increasing over time when $r_t = 0$, and this growth will serve to reduce the marginal utility of the consumption obtained by incremental conversion. This marginal utility will fall faster the larger is η. η is the elasticity of the marginal social utility of per capita consumption. The larger η is, the greater is a country's concern for a more egalitarian distribution of income across generations (see Dasgupta and Heal 1979). If $\eta > 1$, then society

---

[7] The problem is autonomous insofar as time enters Equation (10-4) only through the positive rate of pure time preference.
[8] See Barrett (1992) for a formal proof.
[9] The steady state optimal level of S also depends on δ; see Barrett (1992).

would prefer to smooth out consumption over time, and to do so would conserve more biological diversity than when $\omega = 0$.

When there is population growth as well as technical progress (ω, $\rho > 0$), condition (6) can be rearranged to read $\omega(\eta - 1) > \rho [\alpha (\eta - 1) - (\sigma - 1)]$. If the term in brackets is negative, population growth makes the irreversibility constraint bite even harder. While population growth erodes per capita consumption (for ω given and $r_t = 0$), a larger population benefits more from biodiversity conservation (the function U is utilitarian). If additional wild land conversion leads to little additional consumption (α is "small") and society has a strong conservation ethic ($\sigma > 1$), population growth will favour more conservation. However, population growth can reverse the effect of a large η on the conservation decision. To see this, note that the inequality may also be written as $\rho(\sigma - 1) + (\omega - \alpha\rho)(\eta - 1) > 0$. The term ($\omega - \alpha\rho$) denotes the rate of change in per capita consumption. Hence, a large value for η favours conservation only if per capita consumption is increasing when $r_t = 0$ (i.e., only if $\omega > \alpha\rho$). Population growth works against rising per capita incomes, and so a concern for an equitable distribution of income across generations may not favour conservation when there is population growth.

10.3   Money transfers for global biodiversity conservation[10]

Section 10.2 considered the conservation of domestic biodiversity. But, as noted in the introduction, biodiversity also exhibits global public good characteristics. Hence, there exist incentives for countries to cooperate in conserving global biodiversity. This section develops the framework of the global biodiversity game.

Much of the structure of this game is provided by the Convention on Biological Diversity itself. According to the Convention (Article 20, Paragraph 2):

> The developed country Parties shall provide new and additional financial resources to enable developing country Parties to meet the agreed *full incremental costs* (emphasis added) to them of implementing measures which fulfil the obligations of this Convention . . .

Here, "full incremental costs" can be taken to mean the compensation which ensures that a developing country's well-being does not diminish as a result of the international policy.

Let us suppose that, for nationalistic reasons, developing country j will want to protect S_j^* units of biodiversity unilaterally. Of course, calculating

---

[10] Sections 10.3 and 10.4 are from Barrett (1994a).

S_j^* is no small matter, as we saw in the last section. However, the Convention outlines in Article 8 the policies which should be undertaken unilaterally, and I shall assume that the value of S_j^* can be identified. Now, provided the net benefit of conservation is continuous, a very small increase in the level of protection from S_j^* will diminish j's well-being imperceptibly; the marginal "incremental" cost of conserving one additional unit of biodiversity will be near zero. Successive increases in protection, however, will involve greater sacrifices and hence require higher incremental compensation; the marginal incremental cost of global biodiversity conservation should be increasing.

Let S_j now denote the *incremental* protection of biodiversity – the quantity of biodiversity protected in excess of the nationalistic optimum, S_j^* – and suppose that incremental costs are given by $c_j S_j^2/2$; in other words, suppose that marginal incremental costs increase linearly. These incremental costs represent the minimum compensation required by j to protect S_j incremental units of biodiversity. Under the terms of the Convention, j will protect S_j if compensated by an amount $c_j S_j^2/2$. To keep the analysis simple, assume, innocuously, that all $j = 1, \ldots, N$ developing countries are identical. The developed country parties will want to spread their total incremental cost payment across all developing country parties such that any given level of aggregate incremental biodiversity conservation S, $S = \Sigma_j S_j$, is protected at minimum cost. Given that all developing country parties are assumed to be identical, this means that developed country parties will want to spread the total incremental cost payment out equally, so that the same quantity of incremental biodiversity is conserved by each developing country party. Hence, the cost of incremental conservation to developed country parties is $\Sigma_j c_j S_j^2/2 = c_j S^2/2N$, where $S = \Sigma_j S_j$. Letting $c \equiv c_j/N$, the cost is $cS^2/2$.

Assume as well that all $i = 1, \ldots, n$ developed countries are identical. In contrast to the assumption that developing countries are identical, the assumption that developed countries are identical is not innocuous. Developing country parties have no potential for strategy in this model; they have agreed to be compensated for incremental costs and not to contribute towards global biodiversity conservation. Developed country parties, by contrast, have not agreed how much each will contribute towards global biodiversity conservation. The assumption that developed countries are identical allows us to focus on the free-rider problem, as the bargaining game among identical parties is trivial.[11] However, the assumption also

[11] This is true for a number of solution concepts, such as the Nash bargaining solution and the Shapley value. It is not necessarily true of other concepts. For example, Shapley and Shubik (1969) give examples where the core is empty or very large.

comes at a cost, for the nature of the free-rider problem can change when countries are not identical.

For developed countries, it seems reasonable to assume that the benefit of conserving one unit of incremental global biodiversity, S, will be positive, and that successive increases in S will yield successively smaller incremental benefits. Assume that country i's benefit function is given by

$$B_i(S) = \left(\frac{b}{n}\right)\left(aS - \frac{S^2}{2}\right) \tag{10-7}$$

The parameter b is the slope of the marginal benefit of incremental conservation, summed over all developed countries, while a is just a scale parameter.[12]

Unless developed countries offer compensation to developing countries, no incremental conservation will be forthcoming (S will equal zero). However, any developed country i can make a money transfer of M_i to conserve biodiversity in developing countries. Denote the total of such payments by all developed country parties as $M = \Sigma_i M_i$. We may think of this as the aggregate resources of a Global Biodiversity Conservation Fund. Assuming that the Fund disperses such resources in a cost-effective manner, the quantity of incremental diversity conserved by an aggregate money transfer M will be $M = cS^2/2$. Rearranging, we have

$$S = (2M/c)^{1/2} \tag{10-8}$$

Increases in assistance yield increases in incremental conservation, but each additional dollar of assistance protects less incremental biodiversity than the last.

Substituting Equations (10-8) into (10-7), and taking into account the cost of the financial transfer to country i yields i's net benefit function

$$NB_i(M_i, M_{-i}) = \left(\frac{b}{n}\right)\left[a\left(\frac{2(M_i + M_{-i})}{c}\right)^{1/2} - \frac{(M_i + M_{-i})}{c}\right] - M_i \tag{10-9}$$

where

$$M_{-i} = \sum_{k \neq i} M_k$$

The Nash equilibrium may be easily solved. Maximising Equation (10-9) on the assumption that $\partial M_{-i}/\partial M_i = 0$ yields i's reaction function, $M_i = a^2 b^2 c/2(cn + b)^2 - M_{-i}$. The noncooperative outcome is

---

[12] If the benefit of conserving the very last unit of global biodiversity equals zero, then the parameter a may be interpreted to be the maximum quantity of diversity that can feasibly be saved.

$$M^o = \frac{a^2b^2c}{2(cn + b)^2} \qquad S^o = \frac{ab}{(cn+b)} \qquad (10\text{-}10)$$

The quantity of incremental biodiversity conserved is proportional to the scale parameter a, increasing in b (the slope of the marginal benefit of conservation for all developed countries), decreasing in c (the slope of the marginal incremental cost of biodiversity conservation), and decreasing in n (the number of developed countries). This is just as we would expect.

An important question is how far the noncooperative outcome departs from the full cooperative outcome. The latter is found by choosing M to maximise

$$NB(M) = b[a(2M/c)^{1/2} - M/c] - M \qquad (10\text{-}11)$$

The solution is

$$M^c = a^2b^2c/2(c+b)^2 \qquad S^c = \frac{ab}{(c+b)} \qquad (10\text{-}12)$$

Obviously, $M^o = M^c$ and $S^o = S^c$ when $n = 1$; otherwise, $M^o < M^c$ and $S^o < S^c$.

These two different outcomes are illustrated in Figure 10.1. In the non-cooperative outcome, each country i sets its *own* marginal benefit of conservation equal to the marginal cost of conservation. In the full cooperative outcome, each country i sets the marginal benefit to all developed countries equal to the marginal cost of conservation.[13] The difference in total net benefits between the two outcomes depends in a complicated way on the parameter values. This difference is discussed briefly later in the next section.

10.4   The Biodiversity Convention as a self-enforcing agreement

This section explores the potential for the Convention to improve upon the noncooperative outcome. Towards this, suppose that a subgroup of developed countries cooperate and contribute an amount of money to the Global Biodiversity Conservation Fund which exceeds the amount which that subgroup would contribute under the noncooperative outcome. Suppose further that countries outside this subgroup continue to behave non-cooperatively. Then, so long as the subgroup's choice set is restricted to its Fund contributions, cooperation will be impotent. The reason can be seen by referring back to the reaction function of noncooperators. For every

---

[13] The full cooperative outcome is thus defined in relation to developed countries only.

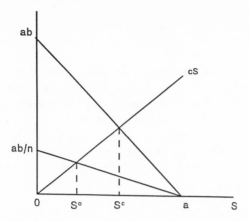

Figure 10.1. The non-co-operative and full co-operative outcomes.

additional $1 contributed by the subgroup, nonmembers will reduce their contributions by $1.[14]

Cooperation can be made more effective if the punishment suffered by free riders is increased. One way of doing this is to play the game repeatedly. In fact, we know from the folk theorem that the full cooperative outcome of the one-shot game can be sustained as a subgame perfect equilibrium of the infinitely repeated game if the rate of discount is sufficiently small.[15] However, the full cooperative outcome will not necessarily be an equilibrium to the biodiversity supergame, even for arbitrarily small discount rates, because the Convention must be self-enforcing and hence must not be vulnerable to renegotiation (Barrett 1994b).

Consider the two-player, repeated prisoners' dilemma game. It is well known that cooperation can be sustained as a subgame perfect equilibrium if both players adopt the "grim" strategy of indefinite reversion to "finking" after any player cheats. However, if one player did cheat, the other might not want to revert to indefinite punishment. After all, that punishment would hurt the country making the punishment as well as the one which that country intended to punish. Both countries would in

[14] Note that this result can be traced to the functional specification of Equation (10-9). While this specification may be plausible, alternative specifications can produce very different results. Barrett (1994b) provides examples where a self-enforcing agreement of any size does not exist, where the self-enforcing agreement consists of exactly two or three countries, and where the self-enforcing agreement consists of from 1 to n countries, depending on the values of key parameters.

[15] Of course, the folk theorem only asserts that any feasible individually rational payoffs can be supported as a Nash equilibrium if the game is repeated infinitely often and players are sufficiently patient. Hence, the full cooperative outcome is only one among many supergame equilibria.

fact have an incentive to renegotiate the agreement. A self-enforcing agreement must clearly be renegotiation proof. However, as we shall soon see, the renegotiation-proof equilibrium of the repeated game may not be able to sustain the full cooperative outcome, even for arbitrarily small discount rates.

Now, supergames assume that players face the same set of feasible actions and per-period payoffs in each of infinitely many periods. But if a country deviates in period t by donating a smaller amount of money to the Fund than stipulated by the agreement, and if reductions in S are irreversible, then the maximal S which can be conserved in period $t+1$ is strictly less than S_t; the game is history-dependent. The inability to allow current feasible actions and payoffs to be influenced by past play is a general problem with supergames. One can consider state-space strategies in an alternative framework, but in doing so one gives up the simplicity of supergames. It is not clear how this assumption influences the results obtained in this paper. On the one hand, irreversibility would make deviators suffer more from any given punishment, and hence serves to deter free-riding. On the other hand, irreversibility would make countries which carry out the punishment suffer more as a result, and hence reduces the magnitude of credible punishments.

Let NB_i now denote country i's *average* payoff in the infinitely repeated game. NB_i is assumed to be given by

$$NB_i(M_i, M_{-i}) = \left(\frac{b}{n}\right)\left[a\left(\frac{2(M_i+M_{-i})}{c}\right)^{1/2} - \frac{(M_i+M_{-i})}{c}\right]$$
$$- M_i - \frac{a^2b^2}{2n(cn+b)} \qquad (10\text{-}13)$$

Equation (10-13) includes a constant term, which is chosen to ensure that each country is guaranteed a payoff of at least zero.[16] The full cooperative outcome yields a payoff to all developed countries of $[a^2b^2c(n - 1)]/[2(c+b)(cn+b)]$. Hence, the set of feasible, individually rational payoffs is:

$$V^* = \left\{v/v_i \geq 0, \sum v_i \leq \frac{a^2b^2c(n - 1)}{2(c + b)(cn + b)}\right\}$$

Following Farrell and Maskin (1989), a payoff vector v is (weakly) renegotiation proof (for discount rates near zero) if payments M^k_k, M^k_z can be chosen to punish country k for cheating such that

---

[16] That is, the constant term is the payoff i receives if NB_i is given by Equation (10-9), and the other $n-1$ countries minimax i by choosing $M_{-i} = 0$. In this case, we know from the reaction function that i will choose $M_{-i} = M^o$. The constant term is thus found by substituting M^o, given in Equation (10-10), for M_i in Equation (10-9).

$$\max_{M_k} \left(\frac{b}{n}\right) \left\{ a\left[2\frac{(M_k + (n-1)M_z^k)}{c}\right]^{1/2} - \right.$$

$$\left. \frac{(M_k + (n-1)M_z^k)}{c} \right\} - M_k - \frac{a^2b^2}{2n(cn+b)} \le v_k \qquad (10\text{-}14)$$

$$\left(\frac{b}{n}\right) \left\{ a\left[2\frac{(M_k^k + (n-1)M_z^k)}{c}\right]^{1/2} - \right. \qquad (10\text{-}15)$$

$$\left. \frac{(M_k^k + (n-1)M_z^k)}{c} \right\} - M_z^k - \frac{a^2b^2}{2n(cn+b)} \ge v_z$$

Condition (10-14) guarantees that punishments M_z^k will deter cheating. Condition (10-15) guarantees that the countries inflicting the punishment have no incentive to renegotiate the agreement.

If we assume that the payoff vector v is Pareto efficient, Conditions (10-14) and (10-15) can be rewritten as

$$(n-1)M_z^k \le v_k \qquad (10\text{-}16)$$

$$\frac{a^2b^2c\,[c(2n-1)+bn]}{2n(c+b)^2(cn+b)} - M_z^k \ge v_z \qquad (10\text{-}17)$$

If country k chooses its minimax money transfer in the punishment phase, then $M_k^k = a^2b^2c/2\,(cn \cdot b)^2$ and $M_z^k = (M^c - M_k^k)/(n-1)$. Substituting, we find that the renegotiation-proof equilibria can support payoffs v_i that lie between the following minimum and maximum values

$$\frac{a^2b^2c[(cn+b)^2 - (c+b)^2]}{2(c+b)^2(cn+b)^2} \le v^{min} \qquad (10\text{-}18)$$

$$\frac{a^2b^2c}{2(c+b)^2(cn+b)} \left(\frac{[c(2n-1)+bn]}{n} - \frac{[(cn+b)^2 - (c+b)^2]}{(n-1)(cn+b)} \right) \ge v^{max}(10\text{-}19)$$

If the full cooperative outcome is to be sustained, we require that each country's payoff in the full cooperative outcome, $v^* = a^2b^2c/2n(c+b)^2$, lie between v^{min} and v^{max}. The maximum number of countries which can sustain the full cooperative outcome, $\bar{n}$, cannot be solved analytically in this case, but simulation analysis reveals that there is a rather simple relationship between $\bar{n}$ and c/b.[17] This is shown in Table 10.1.

---

[17] Note that an analytical solution may be found for a different functional form; see Barrett (1994b). The parameter a is only a scaling factor in the above functional specification, and so does not affect $\bar{n}$.

Table 10.1. *Simulation of maximum number of countries that can sustain the full cooperative outcome*

b	c	c/b	$\bar{n}$	$NB(M^C)$ -$NB(M^O)$
1	1	1	0	20.5
10	1	.1	3	182.3
100	1	.01	7	124.1
1	.01	.01	7	1.2
10	.01	.001	23	0.2
100	.01	.0001	71	0.02
1	.1	.1	3	18.2
10	.1	.01	7	12.4
100	.1	.001	23	1.7
1	10	10	0	4.1
10	10	1	0	204.6
100	10	.1	3	1823.2

Note: The last column in the table gives the gap between the total net benefits in the full co-operative and noncooperative outcomes, assuming $n = 20$ and $a = 10$ (the value of a does not affect the relative values of $NB(M^C) - NB(M^O)$).

According to the table, $\bar{n}$ is smaller the larger is c/b. The intuition behind this finding is roughly as follows. The smaller is b, the smaller is the harm suffered by cooperators in the event of a defection, and hence the smaller is the credible punishment. The larger is c, the larger is the gain to a defection. Hence, the larger is c/b, all else being equal, the smaller is the number of countries which can sustain the full cooperative outcome.

An important question is whether $\bar{n}$ is large when the difference between total net benefits in the full cooperative and noncooperative outcomes is large. This difference is given by

$$NB(M^c) - NB(M^o) = a^2c(n - 1)^2/\{2(c/b + 1)[n^2c/b + 2n + (c/b)^{-1}]\}(10\text{-}20)$$

The difference $NB(M^c) - NB(M^o)$ is increasing in n, as one would expect. It is also increasing in c for c/b constant. Finally, the gap can be shown to be decreasing in c/b if $n(c/b)^2(2nc/b + 2 + n) > 1$ and increasing in c/b if this inequality is reversed. Figure 10.2 plots $NB(M^c) - NB(M^o)$ against c/b for two values of c, assuming $a = 10$ and $n = 20$. As the figure shows, $NB(M^c) - NB(M^o)$ is largest for this simulation when c/b is near 0.05 and c is "large."

The last column in Table 10.1 gives $NB(M^c) - NB(M^o)$ assuming, as in Figure 10.2, $a = 10$ and $n = 20$. When $c/b = .001$ or .0001, Table 10.1 shows that $\bar{n} > n$; the full cooperative outcome can be sustained. But in these cases $NB(M^c) - NB(M^o)$ is quite small. When $b = 100$ and $c = 10$, the gap between the two outcomes is quite large, but in this case the full cooperative outcome cannot be sustained. This suggests that when an agreement is signed by many countries, the agreement may not succeed in

.

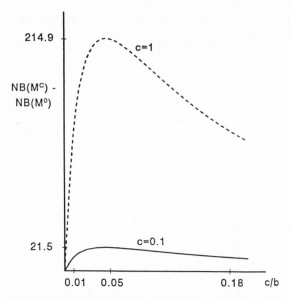

Figure 10.2. The gain to full cooperation.

increasing global net benefits by much compared to the noncooperative outcome (in this case, c/b must be small), and that when net benefits could be increased substantially through cooperation (when c is very large and c/b is, in the above example, near .05), the full cooperative outcome may not be sustainable.[18] This apparent paradox can also be shown to characterise other environmental problems, although it must be emphasised that the results can be sensitive to choice of functional form and model specification (see Barrett 1994b).

10.5   Conclusions

This paper has considered two of the important questions raised by the Convention on Biological Diversity. The first is how countries should incorporate the conservation of biodiversity into development planning. The second is how cooperation to transfer resources for the conservation of incremental biodiversity can be sustained in a self-enforcing international agreement. While the models employed in the paper are highly specialised, and the results may therefore not hold generally, the results

[18] The full cooperative outcome will also not be sustainable when c/b is very large and c is small, but in this case $NB(M^c) - NB(M^o)$ will be small.

that have been obtained are striking. In both cases, the results obtained in the paper have a knife-edge property.

In the first case, it was shown that, depending on parameter values, a country may want to exhaust all of its biodiversity (asymptotically) or conserve more biodiversity than would be optimal on the basis of current marginal valuations. In the second case, it was shown that, depending again on certain parameter values and the model specification, the Convention could only codify the noncooperative outcome or it could sustain the full cooperative outcome where global net benefits are maximised. However, where the agreement can sustain the full cooperative outcome, net benefits will be only slightly larger than in the noncooperative outcome. Where net benefits are much greater in the full cooperative outcome than in the noncooperative outcome, the full cooperative outcome cannot be sustained by a self-enforcing agreement.[19]

[19] This assumes, of course, that n is large.

PART IV
CONCLUSIONS

CHAPTER 11

Unanswered questions

*Charles Perrings, Karl-Göran Mäler, Carl Folke,
C. S. Holling and Bengt-Owe Jansson*

11.1 Biodiversity preservation versus biodiversity conservation

The contributors to this volume have argued that the fundamental goal of biodiversity conservation is not species preservation for its own sake, but the protection of the productive potential of those ecosystems on which human activity depends. This, it has been argued, is a function of the resilience of such ecosystems. Ecosystem resilience has been shown to be a measure of the limits of the local stability of the self-organisation of the system. Hence a system may be said to be resilient with respect to exogenous stress or shocks of a given magnitude if it is able to respond without losing self-organisation. Where species or population deletion jeopardises the resilience of an ecosystem providing essential services, then protection of ecosystem resilience implies species preservation. This is not to say that we should dismiss arguments for species preservation for its own sake. The identification of existence or nonuse value in contingent valuation exercises indicates that people do think in such terms. But it does make it clear that there is both an economically and ecologically sound rationale for ensuring the conservation of species that are not currently in use. More particularly, species which are not now keystone species but may become keystone species under different environmental conditions have insurance value, and this insurance value depends on their contribution to ecosystem resilience.

It is worth underlining the point that where the objective is the maintenance of ecosystem resilience, the appropriate measure of diversity is not genetic distance but what may be termed functional distance. The point was made in Chapter 2 that the vulnerability of particular key structuring processes depends on the number of alternative species that can provide a particular function following perturbation of the system. It also depends on the range of environmental conditions over which such species can provide the function – narrower in more highly structured terrestrial sys-

301

tems and wider in less highly structured coastal and estuarine systems. The least resilient or the most sensitive components of food webs, energy flows and biogeochemical cycles are argued by Holling et al. to be those where the number of species carrying out that function is very small. However, as Costanza et al. observed, there exist resilient estuarine systems where the number of species carrying out the function is small, but the adaptability of each is large. Functional redundancy may have both a species and an unused capacity dimension.

In terms of the economics of biodiversity loss, this perspective on the problem raises a number of interesting questions that are posed but not completely answered in this volume. We conclude with a short discussion of the outstanding research questions, and the challenges they offer to economists and ecologists alike. The challenges are of two sorts. First, there is the methodological challenge of developing a genuinely transdisciplinary approach to the modelling and analysis of ecological-economic systems. While the contributions to this volume have taken our understanding of the biodiversity problem some way forward, economists and ecologists have much to learn about how the analytical techniques in each discipline can best be married. Second, there is the more substantive challenge posed by the complex time behaviour and in-built uncertainty of such ecological-economic systems. The first of these we merely note and pass on to the second.

11.2   Loss of resilience and the dynamics of ecological-economic systems

Every chapter in this volume has emphasised the fact that the dynamics of economic systems are not independent of the dynamics of the ecological systems which constitute their environment. It has also been argued that as economies grow relative to their environment, the dynamics of the jointly determined system are increasingly discontinuous. Neither joint system dynamics nor threshold effects are adequately addressed by existing economic and biological theory. The existence of nonlinear discontinuous dynamics is not, of course, unique to the interaction of economic and ecological systems. Indeed, the theory of complex dynamical systems is a growth area in other disciplines, just as it is in other fields of economics and biology. But it is a general feature of ecological-economic systems, and it is a particular feature of the dynamic effects of biodiversity loss. This influences the construction of the economic problem, the valuation of environmental resources, and the identification of policy options and instruments.

Modelling the time-behaviour of stressed ecological-economic systems involves both the theory of complex system dynamics developed by Kolmogorov and others (cf Li and York 1975), and the theory of dissipative systems developed by Prigogine et al. (cf Nicolis and Prigogine 1977). In ecology, this body of theory has strongly influenced recent research on scale, complexity, stability and resilience, and is beginning to influence the theoretical treatment of the coevolution of species and systems (see the discussion in Chapter 2). In economics there is already a well-established body of theory on the dynamics of complex nonlinear systems (Anderson et al. 1988; Arthur 1992; Brock and Malliaris 1989; Rosser 1990, 1991; Benhabib, 1992) that is motivated by reference to parallel work in epidemiology, biology and ecology (see Brock 1992). In Chapter 6, Perrings and Walker explore some of the issues involved in the construction of an ecological-economic model of a system subject to discontinuous change, but there do not yet exist ecological-economic models of biodiversity loss appealing to these bodies of theory. This remains one of the main areas for biodiversity research in the future.

From a decision perspective, the complex time behaviour of the joint system is significant because of the fundamental uncertainty it implies. This is related to the hierarchical temporal and spatial structure of the ecological system. As Holling et al. point out in Chapter 2, landscapes may be conceptualised as hierarchies, each level of which involves a specific temporal and spatial scale (see also Wiens 1992). The dynamics of each level of the structure will be predictable so long as the biotic potential of that level is consistent with bounds imposed by the remaining levels in the hierarchy. Change in either the structure of environmental constraints or the biotic potential of the level may induce threshold effects that lead to complete alteration in the state of the system. Indeed, this is what is meant by loss of resilience. The problem for decision-making is that such alterations in the state of the system are inherently unpredictable. It is well understood that catastrophic losses occurring with low probability are not amenable to treatment by conventional decision models, and many of the problems created by loss of biodiversity fall into precisely this category. Where the main social costs of biodiversity depletion lie in the loss of ecosystem resilience, they are inherently unpredictable. If the systems affected include those providing essential life support services, it is reasonable to believe that the social costs will be catastrophic. This poses four related challenges to our understanding of the problem that have yet to be satisfactorily resolved. These challenges are: (1) to the theory of decision-making under ignorance; (2) to the valuation of the resources affected by biodiversity loss; (3) to the determination and treatment of the intragenerational and intergenerational equity implications of the loss of diversity; and (4) to the development of appropriate policy responses. The theory of decision-making under ignorance continues to attract attention for other reasons[1] and we merely note that biodiversity loss raises the problem in a particularly acute way. In what follows we consider each of the other challenges in turn.

11.3   Valuation and efficiency

Various contributions to this volume have addressed the problem of biodiversity loss from an efficiency perspective. That is, it has been assumed that the rate of biodiversity depletion is suboptimal because the private users of biotic resources have been confronted by costs of depletion that differ from the social costs of depletion. But this begs the question as to what is the social value of those resources in their current use. The estimation of the private valuation of nonmarketed environmental resources under hypothetical allocations has attracted an enormous amount of attention in recent years. But it is fair to say that there are few convincing attempts to estimate the value to society of the loss of resilience associated with current biodiversity depleting activities. In this volume, the chapter by Turner et al. identifies what they term the "primary" or infrastructural value of ecosystems to be a particular source of difficulty in this respect. This is both because of lack of information about the infrastructural functions of the system, and because of the public good nature of those functions. We will return to the public good properties of ecosystem resilience in Section 11.5.

The challenge to research in this area is to estimate the value of the ecosystem resilience forgone by biodiversity depletion. That is, the challenge is to estimate the stream of benefits forgone as a result of the contraction of the range of environmental conditions in which ecosystems can continue to provide positively valued ecological services. This implies an aggregate "production function" approach in which ecologists and economists should collaborate in the specification of the functional relationship between marketed goods and services and ecosystem resilience. The problem may be likened to that of identifying a premium for the insurance provided by the biodiversity underpinning ecosystem resilience when there is no basis on which to arrive at actuarially fair estimates of the welfare losses associated with loss of resilience.

11.4   Intergenerational and intragenerational equity:
the ethical issues

It is generally recognised that the relation between economic and ecological systems raises questions that go beyond the positivist reach of the natu-

---

[1] See, for example, Katzner (1994).

ral sciences. The conservation of biodiversity poses several such questions in a very acute way. One relates to the fact that biodiversity loss has often irreversible future effects, and so questions the responsibility borne by those living now for the well-being of the members of future generations. This is readily recognisable as a problem of intergenerational equity. A second relates to the fact that loss of ecosystem resilience due to the depletion of biodiversity by any one person affects all those who currently depend on the system, and so bears on the responsibility born each individual for all other individuals in society. This is less obviously recognisable as a problem of intragenerational equity. The ethical issues raised by both questions concern the principle of consumer sovereignty. This principle privileges the rights of the individual not only with respect to the society in which he or she exists, but also with respect to generations yet to come.

The challenge for research into the ecological economics of biodiversity is to identify the rates at which it is ethically "neutral" to discount effects of private behaviour both on other members of the same generation and all members of future generations. As Mäler has argued in this volume, if the growth rate of environmental resources is zero, the decentralised equilibrium interest rate should also be zero, indicating that compensations to future generations should not be discounted. That is, the challenge is to identify the rate at which the current and future effects of biodiversity depletion are naturally "filtered out" or "compensated." If the future effects of biodiversity loss cannot be compensated – as would be the case with irreversible loss of life support functions, say – then the ethically neutral rate of time discount is zero. At present very little is known of the ethically neutral temporal and spatial discount rates. Indeed, the concept of spatial discount rates is only just becoming part of the research agenda. Some biologists have warned that continuation of biodiversity loss at current rates will have effects comparable to a "nuclear winter" by the middle of the next century (Ehrlich 1988), but this cannot be known with certainty. Starret (1991) has shown in the context of an overlapping generations model that the more "essential" are environmental resources as a fixed factor of production, the greater the divergence between private and social rates of discount will tend to be, and the more private decisions will underrepresent future generations (see also Mäler 1991a). It would seem to be important to understand what social rates of discount are themselves consistent with fairness in the treatment both of future generations and others of the present generation.

The purely ethical questions involved are the subject of a continuing debate. At issue is the view that if the existence of the component parts of a system is contingent on the health of the whole, as is the case in an ecological system, then it is not possible to support the "sovereignty" of any one component of the system without at the same time defining the bounds within which sovereignty may be exercised. If this view is accepted, then there is a role for ecological and economic science to identify and protect the bounds within which individual consumers should be free to exercise sovereignty. In reality, consumer sovereignty is hedged about with restrictions designed to protect society from the effects of ill-informed, irrational or malevolent individual behaviour. It remains to establish the restrictions needed to guard against excessive loss of biological diversity.

11.5   Public goods and incremental costs

The last of the challenges to research suggested by the contributions to this volume concerns the problem of public goods. The chapters in Part III all consider the problem of assuring the sustainable use of biodiversity as an environmental public good. At one level it is immaterial whether the public good is conceived as the genetic library or the insurance provided by ecosystem resilience. The problem is the same. How can the structure of payoffs to the decisions of individual users be varied to assure a sustainable outcome. This is the problem being addressed in the work currently being done by the Global Environment Facility on incremental costs. At another level the identification of the public good is critical. The genetic library is a global public good – as is the carbon cycle. The resilience of specific ecosystems, though also a public good in the sense that it is neither rival nor exclusive in consumption, has a more restricted domain.

The significance of this lies in the fact that public goods with a restricted domain are more likely to be located within a single or at least a small number of jurisdictions. This renders the problem of the provision of the public good more tractable. As Barrett shows, the problem with global public goods lies in the fact that the allocation of those goods is in the nature of a noncooperative game, and the scope for achieving a cooperative outcome is sensitive to the difference between the cooperative and noncooperative outcomes.

If biodiversity conservation is important because of the insurance value it offers at the ecosystem level, two things follow. The first is that many more of the benefits of biodiversity conservation are capturable within local jurisdictions. Local authorities have a stronger incentive to seek an optimal allocation of biological resources. The second is that the difference between the cooperative and noncooperative outcomes in respect of the global dimensions of the public good – its information value – will be less. This increases the possibility that agreement between the players (the Biodiversity Convention) may be self-enforcing. The problem for research is that it is not yet clear what is the domain of the relevant public goods.

Nor is it yet clear what is the difference in the payoffs to cooperative and noncooperative behaviour in each case. This last is clearly linked to the issue of valuation discussed earlier. Indeed, it is part of the reason why we identify the valuation of ecosystem resilience as the single most important research problem in the field of ecological economics.

11.6   "Last things"

In the final volume of his sequence of novels, *Strangers and Brothers*, C. P. Snow asked "who would dare to look in the mirror of his future?" The background against which this volume was written is a popular debate in which some of the world's best known biologists have warned that if humanity does not slow the rate at which species are being driven to extinction the mirror of our collective future will be blank. The vision of the authors contributing to this volume is less apocalyptic, perhaps. But it too gives cause for concern. The problem is not that private decisions on the use of biological resources involve externalities. Externalities are, after all, a feature of almost every transaction undertaken. It is that those externalities may involve irreversible loss of resilience in ecological systems that provide essential life support services. Loss of resilience implies a narrowing of the range of environmental conditions in which the system concerned can maintain its productive potential. This is equivalent to a weakening of the insurance offered by what ecologists have termed "functional redundancy" in the system: its ability to maintain productivity under a range of environmental conditions. Since all the general circulation models (of climate change) predict an increase in the range of environmental conditions within which economic and ecological systems will have to function in the future, the loss of resilience in key ecosystems must be a matter for concern.

This being so it is important to understand the processes involved in biodiversity loss, and that has been our purpose in this volume. The ecological reviews of the issue in Chapters 2 and 3 showed how varied may be the links between functional and species diversity, and between functional diversity and ecosystem resilience. Understanding the economic significance of this requires collaborative work at the ecosystem level between the natural and the social sciences, and Part II of this volume includes three examples of such work. The divergent approaches adopted in each case reflect the fact that such collaborative work is still in its infancy, but it is clear that an ecological-economic model substantially alters our perception of the "environmental conditions" with respect to which the system may be resilient. As Chapter 6 showed, economic shocks test the resilience of ecological systems (through the management regime) in much the same way as do fluctuations in precipitation or temperature.

The policy implications of an ecological-economic approach to the problem of biodiversity loss have yet to be spelled out clearly, though several options are explored in Perrings et al. (1994). What is becoming clear, though, is that the sustainability of economic development implies ecological stabilisation: the maintenance of the productive potential of ecosystems supplying essential ecological services either by the containment of stress levels or by the promotion of ecosystem resilience through biodiversity conservation. As with other policy objectives, ecological stabilisation requires the development of an appropriate set of incentives, and these are only now being considered. Much of the international effort on incentives has gone into the development of instruments designed to assure efficiency in the allocation of environmental resources within safe bounds, including harvesting or hunting quota, emission caps or "bubbles," time-limited rights or "seasons." The bounds may be set by scientists, but are more often rooted in custom or generated by some political process. The setting of bounds, and the development of the corresponding incentive structures is one of the areas in which collaborative work between ecologists and economists is most needed. Indeed, as humanity presses ever more closely against the thresholds of ecosystem resilience in one biome after another, it is likely that such work will become of critical importance.

The implications of crossing thresholds of ecosystem resilience have always been highly uncertain, but that uncertainty mattered little so long as those thresholds were some way off. As they get nearer, however, uncertainty about the costs and benefits of stepping across thresholds has at least focused attention on the true extent of our collective ignorance. We are beginning to appreciate the implications of knowledge that stops at the threshold. We now understand that it is unsatisfactory to concentrate our efforts on marginal increments in knowledge of the environmental costs of economic activity when the only thing we are sure about is that those environmental costs are not continuous. As we near the thresholds of resilience of key ecosystems we must not only dare to look into the mirror of our future, we must make sure that we understand what we see.

References

Abel N.O.J. 1990. Destocking communal pastures in Southern Africa: Is it worth it? Paper prepared for the Technical Meeting on Savannah Development and Pasture Production, Commonwealth Secretariat and the Overseas Development Institute, Woburn, November 1990.

Allen T.F.H. and Starr T.B. 1982. *Hierarchy: Perspectives for ecological complexity.* University of Chicago Press, Chicago.

Amelung T. 1991. Tropical Deforestation as an International Economic Problem. Paper presented at the Egon-Sohmen Foundation Conference on Economic Evolution and Environmental Concerns, Linz, Austria, 30–31 August.

Amelung T. and Diehl M. 1991. Deforestation of Tropical Rainforests: Economic Causes and Impact on Development, *Kieler Studien 241.* Tubingen, Mohr, Germany.

Anderson P., Arrow K. and Pines D. 1988. The economy as an evolving complex system, *Santa Fe Institute Studies in the Sciences of Complexity V.* Addison Wesley, Redwood City, CA.

Andreàsson-Gren I.M. 1991. "Costs for nitrogen source reduction in a eutrophicated bay in Sweden" in Folke C. and Kaberger T. eds., *Linking the natural environment and the economy: Essays from the eco-eco group.* Kluwer Academic Publishers, Dordrecht: 173–188.

Arrow K.J. and Fisher A.C. 1974. Environmental preservation, uncertainty, and irreversibility, *Quarterly Journal of Economics, 88, 2*: 312–319.

Arthur B. 1990. Positive feedback in the economy, *Scientific American 26:* 92–99.

Arthur B. 1992. *On learning and adaptation in the economy,* Food research Institute, Stanford University, Stanford.

Austin M.P., and Williams O. 1988. Influence of climate and community composition on the population demography of pasture species in semi-arid Australia, *Vegetation, 77:* 43–49.

Ayres R. and Kneese A. V. 1969. Production, Consumption and Externalities, *American Economic Review, 59* 282–297.

Barbier E.B. 1988. *Sustainable agriculture and the resource poor: policy issues and options,* LEEC Paper 88-102, London Environmental Economics Centre, London.

Barbier E.B. 1989a. *Economic Evaluation of Tropical Wetland Resources: Applications in Central America.* LEEC, London.

Barbier E.B. 1989b. The Economic Value of Ecosystems: 1 - Tropical Wetlands, *LEEC Gatekeeper Series 89-101,* LEEC, London.

Barbier E.B. 1989c. The contribution of environmental and resource economics to an economics of sustainable development, *Development and Change 20:* 429–459.

Barbier E.B. 1994. Valuing environmental functions: Tropical wetlands, *Land Economics 70(2):* 155–173.

Barbier E.B., Bockstael N., Burgess J.C. and Strand I. 1994a. The linkages between the timber trade and tropical deforestation, *World Economics,* in press.

Barbier E.B., Burgess J.C., and Markandya A. 1991. The Economics of Tropical Deforestation, *Ambio 20:* 55–58.

Barbier E.B., Burgess J.C., Aylward B.A., Bishop J.T. and Bann C. 1994b. *The Economics of the Tropical Timber Trade,* Earthscan Publications, London.

Barbier E., Burgess J., Swanson T., and Pearce D. 1990. *Elephants, Economics and Ivory,* Earthscan Publications, London.

Barbier E.B. and Markandya A. 1990. The Conditions for Achieving Environmentally Sustainable Development, *European Economic Review 34:* 659–669.

Barbier E.B., Markandya A. and Pearce D.W. 1990. Environmental Sustainability and Cost-Benefit Analysis, *Environment and Planning 22:* 1269–1266.

Barbier E.B. and Rauscher M. 1994. Trade, Tropical Deforestation and Policy Interventions, *Environmental and Resource Economics 3:* 35–50.

Barnes D.L. 1979. Cattle ranching in the semi-arid savannas of east and southern Africa. in Walker B.H. ed. *Management of Semi-arid Ecosystems.* Elsevier, Amsterdam, 9–54.

Barrett S. 1990. The Problem of Global Environmental Protection, *Oxford Review of Economic Policy 6:* 68–79.

Barrett S. 1992. Economic Growth and Environmental Preservation, *Journal of Environmental Economics and Management 23:* 289–300.

Barrett S. 1994b. Self-Enforcing International Environmental Agreements, *Oxford Economic Papers 46*: 878–894.

Barrett S. 1993. Optimal Economic Growth and the Conservation of Biological Diversity, in Barbier E. ed. *Economics and Ecology,* Chapman & Hall, London: 130–145.

Barrett S. 1994a. The Biodiversity Supergame, *Environmental and Resource Economics,* 4:111–112.

Bateman I.J. and Turner R.K. 1992. The Contingent Valuation Method: A Theoretical and Methodological Assessment, *CSERGE Working Paper GEC 92-18,* Centre for Social and Economics Research on the Global Environment, UCL/UEA, London and Norwich.

Bateman I.J., Willis K.G., Garrod G.D., Doktor P., Langford I. and Turner R.K. 1992. *Recreation and Environmental Preservation Value of the Norfolk Broads: A Contingent Valuation Study,* Report to the National Rivers Authority, Environmental Appraisal Group, University of East Anglia.

Baumol W. and Bradford D. 1972. Detrimental externalities and non-convexity of the production set, *Economica 39:* 160–176.

Bayley S. E., Schindler D.W., Parker B.R., Stainton M.P. and Beaty K.G. 1993. Effect of forest fire and drought on acidity of a base-poor boreal forest stream: similarities between climatic warming and acidic precipitation, Biogeochem, Mimeo.

Bayley S., Stotts, V.D., Springer, P.F. and Steenis, J. 1978. Changes in submerged aquatic macrophyte populations at the head of the Chesapeake Bay, 1958–1975. *Estuaries* 1: 74–85.

Begon M., Harper J.L. and Townsend C.R. 1987. *Ecology: Individuals, Populations and Communities,* Blackwell, Oxford.

Behnke R.H. 1985. Measuring the benefits of subsistence versus commercial livestock production in Africa, *Agricultural Systems 16:* 109–135.

Benhabib J. ed. 1992. *Cycles and Chaos in Economic Equilibrium,* Princeton University Press, Princeton.

Bergström J.C., Stoll J. R.T., Tire H.P., and Wright V.L. 1990. Economic value of wetlands-based recreation, *Ecological Economics 2:* 129–147.

Berkes F. and Folke C. 1992. A Systems Perspective on the Interrelations between natural, human-made and cultural capital, *Ecological Economics 5:* 1–8.

Binkley C.S. and Vincent J.R. 1991. Forest Based Industrialisation: A Dynamic Perspective, *HIID Development Discussion Paper No. 389,* Harvard University, Cambridge, Massachusetts.

Biot Y. 1990. How long can high stocking densities be sustained? Paper prepared for the Technical Meeting (November) on Savannah Development and Pasture Production, Commonwealth Secretariat and the Overseas Development Institute, Woburn.

Blanchard O. and Fischer S. 1988. *Lectures on Macroeconomics,* The MIT Press, Cambridge, Massachusetts.

Bockstael N.E. and McConnell K.E. 1983. Welfare measurement in the household production framework, *American Economic Review 73:* 806–814.

Boesch D.F. 1974. Diversity, stability and response to human disturbance in estuarine ecosystems. In *Proceedings of 1st International Congress, Ecological Center for Agricultural Publications and Documents,* The Hague, Netherlands: 109–114.

Bormann F.H. and Likens G.E. 1981. *Patterns and Process in a Forested Ecosystem,* Springer-Verlag, New York.

Braithwaite W. and Walker B.H. 1989. Forest conservation in Australia. In *Proceedingss of the 14th Commonwealth Forestry Conference,* Kuala Lumpur, Malaysia.

Brock W.A. 1992. Pathways to randomness in the economy: emergent nonlinearity and chaos in economics and finance, SSRI Working Paper, University of Wisconsin, Madison.

Brock W.A. and Malliaris A.G. 1989. *Differential Equations, Stability and Chaos in Dynamic Economics,* North Holland, Amsterdam.

Bromley D. 1991. *Environment and Economy: Property Rights and Public Policy,* Blackwell, Oxford.

Brown G.M. 1990. Valuation of Genetic Resources. In Orians G.H., Brown G.M., Kunin W.E. and Swierbinski J.E. eds. *The Preservation and Valuation of Biological Resources,* University of Washington Press, Seattle: 203–228.

Brown J. H. and Maurer B.A. 1989. Macroecology: the division of food and space among species on continents, *Science 243:* 1145–1150.

Bruce J. 1988. A perspective on indigenous land tenure systems and land concentration, in Down R.E. and Reyna S. eds. *Land and Society in Contemporary Africa,* University Press of New England, Hanover, NH.

Buongiorno J. and Manuring T. 1992. Predicted Effects of an Import Tax in the

European Community on the International Trade of Tropical Timbers, Department of Forestry, University of Wisonsin, Madison, Mimeo.

Burgess J.C. 1992. The Impact of Wildlife Trade on Endangered Species, *LEEC Discussion Paper, 92*–102, London Environmental Economics Centre, London.

Carpenter S.R. ed. 1988. *Complex Interactions in Lake Communities,* Springer-Verlag, New York.

Carpenter S.R. and Leavitt P.R. 1991. Temporal variation in paleoimmunological record arising from a tropic cascade, *Ecology 72:* 277–285.

Clarke C. 1990. *Mathematical Bioeconomics: The Optimal Management of Renewable Resources,* New York, John Wiley and Sons.

Clark W.C., Jones D.D. and Holling C.S. 1979. Lessons for ecological policy design: a case study of ecosystem management, *Ecological Modelling 7:* 1–53.

Cleaver K. 1988. The use of price policy to stimulate agricultural growth in sub-Saharan Africa, paper presented to the 8th Agricultural Sector Symposium on Trade, Aid, and Policy Reform for Agriculture.

Clements F.E. 1916. Plant succession: an analysis of the development of vegetation, *Carnegie Institute of Washington Publication 242:* 1–512.

Cohen J. 1991. Tropic topology, *Science 251:* 686–687.

Common M. and Perrings C. 1992. Towards an Ecological Economics of Sustainability, *Ecological Economics 6:* 7–34.

Conrad J. and Clark C. 1987. *Natural Resource Economics,* Cambridge University Press, Cambridge.

Constantino L.F. 1990. *On the Efficiency of Indonesia's Sawmilling and Plymilling Industries.* D.G. of Forest Utilisation, Ministry of Forestry, Government of Indonesia and FAO Jakarta.

Conway G.R. 1987. The properties of agroecosystems, *Agricultural Systems 24:* 95–117.

Conway G.R. and Barbier E.B. 1990. *After the green revolution: Sustainable Agriculture for Development,* Earthscan, London.

Costanza R. and Daly H.E. 1992. Natural capital and sustainable development. *Conservation Biology 6:* 37–46.

Costanza R., Farber C.S. and Maxwell J. 1989. Valuation and management of wetland ecosystems, *Ecological Economics No. 1:* 335–361.

Cousins B. 1987. *A survey of current grazing schemes in the communal lands of Zimbabwe.* UZ. Centre for Applied Social Sciences, Harare.

Cumberland J.H. ed. 1988. *Proceedings, Third Annual Conference on the Economics of Chesapeake Bay Management, Annapolis, Maryland.* Bureau of Business and Economic Research, University of Maryland.

Cumberland J.H. ed. 1989. *Proceedings, Fourth Annual Conference on the Economics of Chesapeake Bay Management, Baltimore, Maryland.* Bureau of Business and Economic Research, University of Maryland.

Cumberland J.H. ed. 1990. *Proceedings, Fifth Annual Conference on the Economics of Chesapeake Bay Management, Annapolis, Maryland.* Bureau of Business and Economic Research, University of Maryland.

Cyert R. and DeGroot M. 1987. Sequential Investment Decisions, in Cyert R. and DeGroot M. *Bayesian Analysis and Uncertainty in Economic Theory,* Chapman and Hall, London.

Daly H.E. and Cobb J.B. 1989. *For the Common Good: Redirecting the Economy*

Towards Community, the Environment and a Sustainable Future, Beacon Press, Boston.

d'Arge R., R. Ayres, and A.V. Kneese. 1970. *Economics and the environment: A Materials Balance Approach,* The Johns Hopkins Press, Baltimore.

Darnell R.M. 1961. Trophic spectrum of an estuarine community, based on studies of Lake Pontchartrain, Louisiana, *Ecology 42:* 553–568.

Dasgupta P. 1982. *The Control of Resources,* Basil Blackwell Publishers Ltd., Oxford.

Dasgupta P. 1990. Commentary on Brown G.M. Valuation of Genetic Resources. In Orians G.H., Brown G.M., Kunin W.E. and Swierbinski J.E. eds. *The Preservation and Valuation of Biological Resources,* University of Washington Press, Seattle: 229–239.

Dasgupta P. 1991. The environment as a commodity. In Blasi P. and Zamagni S. eds. *Man-Environment and Development: Towards a Global Approach,* Nova Spes International Foundation Press, Rome: 149–180.

Dasgupta P. and Heal G. 1979. *Economic Theory and Exhaustible Resources,* Cambridge University Press, Cambridge.

Dasgupta P. and Mäler K.G. 1991. The Environment and Emerging Development Issues, in *Proceedings of the World Bank Annual Conference on Development Economics 1990,* Washington DC, World Bank: 101–131.

Dasgupta P. and Mäler K.G. 1995. Poverty, Institutions, and the Environmental Resource Base. In Behrman J. and Srinivasan T.N. eds. *Handbook of Development Economics vol. 3,* North-Holland, Amsterdam: in press.

David P. 1985. Clio and the Economics of QWERTY, *American Economic Review, Papers and Proceedings, 75:* 332–337.

Davis M.B. 1981. Quaternary history and the stability of forest communities. In West D.C., Shugart H.H. and Botkin D.B. eds., *Forest Succession: Concepts and Application,* Springer-Verlag, New York: 132–153.

Davis M.B. 1986. Climatic instability, time lags and community disequilibrium. In Diamond J. and Case T. eds., *Community ecology,* Harper and Row, New York: 269–284.

Day J.W. Jr., Hall C.A.S., Kemp W.M. and Yanez-Arancibia A. 1989. *Estuarine Ecology,* John Wiley, New York.

De Angelis D.L. 1980. Energy flow, nutrient cycling and ecosystem resilience, *Ecology 61:* 764–771.

Delcourt H.R., Delcourt P.A. and Webb T.I. 1983. Dynamic plant ecology: the spectrum of vegetational change in space and time. *Quaternary Science Reviews 1:* 153–175.

Dembner S. 1991. Provisional Data from the Forest Resources Assessment 1990 Project, *UNASYLVA 42:*40–44.

Di Castri F. 1987. The Evolution of Terrestrial Ecosystems, In Ravera O. ed. *Ecological Assessment of Environmental Degradation, Pollution and Recovery,* Elsevier, Amsterdam.

Dixit A. 1992. Investment and Hysteresis, *The Journal of Economic Perspectives 6:* 7–32.

Dixon J.A. 1989. Valuation of Mangroves, *Tropical Coastal Area Management* 4: 1-6.

Dixon J.A., James D.E. and Sherman P.B. 1989. *The Economics of Dryland Management,* Earthscan, London.

Dublin H.T., Sinclair A.R.E. and McGlade J. 1990. Elephants and fire as causes of multiple stable states in the Serengeti-mara woodlands, *Journal of Animal Ecology 59:* 1147–1164.

Ehrlich P.R. 1988. The loss of diversity: causes and consequences. In Wilson E.O. ed., *Biodiversity,* National Academy Press, Washington: 21–27.

Ehrlich P.R. and Ehrlich A.H. 1991. The value of biodiversity, Stanford University Department of Biological Sciences, Stanford, Mimeo.

Ehrlich P.R. and Ehrlich A.H. 1992. The Value of Biodiversity, *Ambio* 21: 219–226.

Ehrlich P.R. and Mooney H.A. 1983. Extinction, substitution, and ecosystem services, *BioScience, 33:* 248–254.

Elliott J.E. 1980. Marx and Schumpeter on capitalism's creative destruction: a comparative restatement, *Quarterly Journal of Economics 95:* 46–58.

Ewel K.C. 1991. Diversity in wetlands. *Evolutionary trends in plants 5:* 90–92.

Ewel K.C. and Odum H.T. 1984. eds. *Cypress Swamps,* University Presses of Florida, Gainesville.

Farrell J. and Maskin E. 1989. Renegotiation in Repeated Games, *Games and Economic Behavior 1:* 327–360.

Farrell T.M., Bracher D. and Roughgarden J. 1991. Cross-shelf transport causes recruitment to intertidal populations in central California, *Limnology and Oceanography 36:* 279–288.

Fee E.J. and Hecky R.E. 1994. The effect of lake size on nutrient fluxes to the mixed-layer during summer stratification, *Canadian Journal of Fisheries and Aquatic Science 50:* forthcoming.

Fenchel T. and Blackburn T.H. 1979. *Bacteria and Mineral Cycling,* Academic Press, London.

Ferguson I.S. and Muñoz-Reyes Navarro J. 1992. *Working Document for ITTO Expert Panel on Resources Needed by Producer Countries to Achieve Sustainable Management by the Year 2000,* Yokohoma, ITTO.

Fiering M.B. 1982. Alternative indices of resilience, *Water Resources Research 18:* 33–39.

Fisher A.C. and Hanemann W.M. 1986. Option values and the extinction of species, *Advances in Applied Micro Economics 4:* 169–190.

Folke C. 1991a. Socioeconomic dependence on the life-supporting environment. In Folke C. and Kaberger T. (eds.) *Linking the natural environment and the economy: Essays from the eco-eco group.* Kluwer Academic Publishers, Dordrecht: 77–94.

Folke C. 1991b. The societal value of wetland life support. In Folke C. and Kaberger T. (eds.) *Linking the natural environment and the economy: Essays from the eco-eco group.* Kluwer Academic Publishers, Dordrecht: 141–171.

Folke C. and Berkes F. 1992. Cultural capital and natural capital interrelations, *Beijer Discussion Paper Series No. 8.* Beijer International Institute of Ecological Economics, The Royal Swedish Academy of Sciences, Stockholm.

Freeman A.M. 1985. Methods for assessing the benefits of environmental programs, in Kneese A.V. and Sweeney J.L. eds. *Handbook of Natural Resource and Energy Economics I,* North-Holland, Amsterdam: 223–270.

Friedel M.H. 1991. Range condition assessment and the concept of threshold, A viewpoint, *Journal of Range Management 44:* 422–426.

Funderburk S.L., Mihursky, J.A., Jordan, S.J. and Riley, D. (eds.) 1991. *Habitat requirements for Chesapeake Bay living resources.* Chesapeake Research Consortium, Inc. Solomons, Maryland.

Gillis M. 1990. Forest Incentive Policies, Paper prepared for the World Bank Forest Policy Paper, The World Bank, Washington, DC.

Goodin R. 1982. Discounting discounting, *Journal of Public Policy 2:* 53–71.

Goodwin R.M. 1990. *Chaotic Economic Dynamics,* Clarendon, Oxford.

Gosselink J.G. and Turner R.E. 1978. The Role of Hydrology in Freshwater Wetland Ecosystems. In Good, R.E., Whigham D.F. and Simpson R.L. eds. *Freshwater Wetlands: Ecological Processes and Management Potential,* Academic Press, New York: 63–78.

Graetz R.D. 1986. A comparative study of sheep grazing in a semi-arid saltbush pasture in two condition classes, *Australian Rangeland Journal 8:* 46–56.

Grainger A. 1990. *The Threatening Desert – Controlling Desertification,* Earthscan, London.

Grassle J.F. 1991. Deep-sea benthic biodiversity. *Bioscience 41:* 464–469.

Gren I.M. 1992. Benefits from restoring wetlands for nitrogen abatement: a case study of Gotland, *Beijer Discussion Paper Series 14,* The Beijer International Institute of Ecological Economics, Stockholm.

Grimm E. C. 1983. Chronology and dynamics of vegetation change in the prairie-woodland region of southern Minnesota, *New Phytologist 93:* 311–335.

Grow G.S. 1987. Polar Bears Have Solar Hairs, *Christian Science Monitor* November 2–8: 24.

Gunderson L. H. 1992. *Spatial and Temporal Hierarchies in the Everglades Ecosystem with Implications to Water Deliveries to Everglades National Park.* Doctoral thesis, University of Florida, Gainesville, Florida.

Hanneman M. 1989. Information and the concept of option value, *Journal of Environmental Economics and Resource Management 16:* 23–37.

Hardin G. 1968. The tragedy of the commons, *Science 162:* 1243–1248.

Hartwick J. 1990. Natural resources, national accounting and economic depreciation, *Journal of Public Economics 43:* 291–304

Hartwick J. 1992. Deforestation and national accounting, *Environmental & Resource Economics 2:* 513–521

Heinselman M.L. 1973. Fire in the virgin forests of the Boundry Waters Canoe Area, *Quaternary Research 3:* 329–382.

Heinselman M.L. 1981. Fire and succession in the conifer forest of Northern America. In *Forest Succession: Concepts and Application,* Springer-Verlag, New York: 374–405.

Henriksen K. and Kemp W.M. 1988. Nitrification in estuarine and coastal marine sediments: methods, patterns, and regulating factors. In Blackburn T.H. and Sorensen J. eds. *Nitrogen Cycling in Coastal Marine Environments,* John Wiley, New York: 207–249.

Henry C. 1974a. Investment decisions under uncertainty: the irreversibility effect, *American Economic Review, 64:* 1006–12.

Henry C. 1974b. Option Values in the Economics of Irreplaceable Assets, Review of Economic Studies Symposium on Economics of Exhaustible Resources, 89–104.

Hicks J.R. 1946. *Value and Capital,* OUP, Oxford.

Hobbs R.J. and Atkins L. 1988. Effect of disturbance and nutrient addition on native and introduced annuals in plant communities in the Western Australian wheatbelt, *Australian Journal of Ecology 13:* 171–179.

Holling C.S. 1973. Resilience and stability of ecological systems, *Annual Review of Ecology and Systematics 4:* 1–23.

Holling C.S. 1980. Forest insects, forest fires and resilience. In Mooney H., Bonnicksen J.M., Christensen N.L., Lotan G.E. and Reiners W.A. eds. *Fire Regimes and Ecosystem Properties.* USDA Forest Service General Technical Report WO–26, USDA, Washington DC.

Holling C.S. 1986. Resilience of ecosystems, local surprise and global change. In Clark W.C. and Munn R.E. eds. *Sustainable Development of the Biosphere,* Cambridge University Press, Cambridge: 292–317.

Holling C.S. 1987. Simplifying the complex, the paradigms of ecological function and structure, *European Journal of Operational Research 30:* 139–146.

Holling C.S. 1988. Temperate forest insect outbreaks, tropical deforestation and migratory birds, *Memoirs of the Entomological Society of Canada 146:* 21–32.

Holling C.S. 1992a. Cross-scale morphology geometry and dynamics of ecosystems, *Ecological Monographs 62:* 447–502.

Holling C.S. 1992b. The role of forest insects in structuring the boreal landscape, In Shugart H.H., Leemans R. and Bonan G.B. eds. *A Systems Analysis of the Global Boreal Forest,* Cambridge University Press, Cambridge: 170–191.

Holling C.S., Jones D.D., and Clark W.C. 1977. Ecological policy design: a case study of forest and pest management. In Norton G.A. and Holling C.S. eds. *Proceedings of a Conference on Pest Management,* October 1976, Laxenburg, Austria, IIASA CP-77-6:13–90.

Howarth R.B. 1991. Intergenerational Competitive Equilibria Under Technological Uncertainty and an Exhaustible Resource Constraint, *Journal of Environmental Economics and Management 21:* 225–243.

Hyde W.F., Newman D.H., and Sedjo R.A. 1991. Forest Economics and Policy Analysis, An Overview, *World Bank Discussion Paper No. 134,* The World Bank, Washington DC.

International Tropical Timber Council (ITTC) 1992a. *Report on the Eighth Committee of Parties to the Convention on International Trade in Endangered Species,* Kyoto, 1–13 March 1992. PCM(X)/7, 10th Session, Yaoundé, Cameroon, 6–14 May.

International Tropical Timber Council (ITTC) 1992b. *Report on Preparations for the 1992 United Nations Conference on Environment and Development.* ITTC(-XII)/8, 10th Session, Yaoundé, Cameroon, 6–14 May.

International Tropical Timber Council (ITTC) 1992c. Report on Preparations for the 1992 United Nations Conference on Environment and Development. ITTC(XII)/8, 10th Session, Yaoundé, Cameroon, 6–14 May.

Jackson J. 1991. Adaptation and diversity of reef corals: patterns result from species differences in resource use and life histories and from disturbances, *Bioscience 41:* 475–482.

Jamal V. 1983. Nomads and farmers, incomes and poverty in rural Somalia. In Ghai D. and Radwan S. eds. *Agrarian Policies and Rural Poverty in Africa,* Geneva, ILO: 281–311.

Johansson P.O. 1987. *The Economic Theory and Measurement of Environmental Benefits,* Cambridge University Press, Cambridge.

Johnston C.A. 1991. Sediment and nutrient retention by freshwater wetlands: Effects on surface water quality. *Critical reviews in environmental control 21:* 491–565.

Jones P.W., Speir H.J., Butowski N.H., O'Reilly R., Gillingham L., and Smoller E. 1990. Chesapeake Bay Fisheries: Status, Trends, Priorities and Data Needs. ASMF-MD-0048. Versar, Inc. Columbia, Maryland.

Katzner D. 1994. *Time, Ignorance and Uncertainty in Economic Models,* Ann Arbor, University of Michigan Press, forthcoming.

Kelly J. and Levin S. 1986. A comparison of aquatic and terrestrial nutrient cycling and production processes in natural ecosystems. In Kullenberg G. Reidel D. ed. *The Role of the Oceans as a Waste Disposal Option,* Dordrecht, Press,The Netherlands: 157–176.

Kelly R.D. and Walker B.H. 1976. The effects of different forms of land use on the ecology of a semi-arid region in south-eastern Rhodesia, *Ecology 64:* 553–576.

Kemp W.M. and Boynton W.R. 1984. Spatial and temporal coupling of nutrient inputs to estuarine primary production: The role of particulate transport and decomposition, *Bulletin of Marine Science.* 35: 522–535.

Kemp W.M., Boynton W.R., Twilley R.R., Stevenson J.C., and Means J.C. 1983. The decline of submerged vascular plants in upper Chesapeake Bay: Summary of results concerning possible causes, *Marine Technological Society Journal* 17: 78–89.

Kemp W.M. and Mitsch W.J. 1979. Turbulence and phytoplankton diversity: a general model of the paradox of plankton, *Ecological Modelling 7:* 201–222.

Kemp W.M., Sampou P.A., Caffrey J.M., Mayer M., Henriksen K., and Boynton W.R. 1990. Ammonium recycling versus denitrification in Chesapeake Bay sediments, *Limnology and Oceanography 35:* 1545–1563.

Knoop W.T. and Walker B.H. 1985. Interactions of woody and herbaceous vegetation in a southern African savanna, *Journal of Ecology 73:* 235–253.

Kolstad C.D. and Braden J.B. 1991. Environmental demand theory, in Braden J.B. and Kolstad C.D. eds. *Measuring the Demand for Environmental Quality,* North Holland, Amsterdam: 17–40.

Koopmans T.C. 1965. On the Concept of Optimal Economic Growth, *Pontificiae Academiae Scientiarum Scripta Varia 28:* 225–288.

Krutilla J.V. 1967. Conservation reconsidered, *American Economic Review 57:* 777–786.

Krutilla J.V. and Fisher A.C. 1985. *The Economics of Natural Environments: Studies in the Valuation of Commodity and Amenity Resources,* Resources for the Future, Washington DC.

Kusler J.A. and Kentula M.E. eds. 1990. *Wetland creation and restoration, The status of the science,* Island Press, Washington D.C.

Laycock W.A. 1991. Stable states and thresholds of range condition on North American rangelands: A viewpoint, *Journal of Range Management 44:* 427–433.

Lee R. 1991. Comment: The Second Tragedy of the Commons. In Davis K. and Bernstam M. eds. *Resources, Environment and Population: Present Knowledge and Future Options,* Princeton University Press, Princeton.

Levin S.A. 1992. The problem of pattern and scale in ecology, *Ecology 73:* 1943–1967.

Li T. and York J. 1975. Period three implies chaos, *American Mathematical Monthly 82:* 985–992.

Likens G.E. 1992. *The Ecosystem Approach: Its Use and Abuse.* Ecology Institute, Oldenburg/Luhe.

LMER Coordination Committee 1992. Understanding Changes in Coastal Environments: the Land Margin Ecosystem Research Program, *EOS 73:* 481–485.

Lovelock J. 1988. *The Ages of Gaia.* W.W. Norton, New York.

Ludwig D., Jones D.D., and Holling C.S. 1978. Qualitative analysis of insect outbreak systems: the spruce budworm and forest. *Journal of Animal Ecology 44:* 315–332.

MacArthur R. and Wilson E. 1967. *The Theory of Island Biogeography,* Princeton University Press, Princeton.

McNaughton S.J. 1985. Ecology of a grazing ecosystem: the Serengeti, *Ecological Monographs 55:* 259–294.

McNaughton S.J., Guess, R.W., and Seagle S.W. 1988. Large mammals and process dynamics in African ecosystems, *BioScience 38:* 791–800.

Mäler K.G. 1974. *Environmental Economics – A Theoretical Inquiry,* Johns Hopkins Press, Baltimore.

Mäler K.G. 1985. Welfare Economics and the Environment. In Kneese A.V. and Sweeney J.L. eds. *Handbook of Natural Resource and Energy Economics,* North Holland, I, Amsterdam: 3–60.

Mäler K.G. 1990. International environmental problems, *Oxford Review of Economic Policy 6:* 80–108.

Mäler K.G. 1991a. *Sustainable Development,* Stockholm School of Economics, Stockholm, Mimeo.

Mäler K.G. 1991b. National accounts and environmental resources, *Environmental and Resource Economics 1:* 1–15.

Mäler K.G. 1992. Multiple Use of Environmental Resources: A Household Production Function Approach to Valuing Resources, *Beijer Discussion Papers No 4.* Beijer Institute, Royal Swedish Academy of Sciences, Stockholm.

Maltby E. 1986. *Waterlogged Wealth: Why Waste the World's Wet Places?* Earthscan, London.

Mann K.H. 1982. Ecology of Coastal Waters, University of California Press, Berkeley.

Markandya A. 1991. Technology, Environment and Employment: A Survey, World Employment Programme Working Paper, ILO Geneva.

May R.M., Beddington J.R., Clark C.W., Holt S.J., and Laws R.M. 1979. Management of multispecies fisheries, *Science 205:* 267–277.

Miller K.R., Furtado J., de Klemm C., McNeely J.A., Myers N., Soule M.E., and Trexton M.C. 1985. Maintaining Biological Diversity: The Key Factor for a Sustainable Society, International Union for Conservation of Nature, Gland.

Miller J. and Lad F. 1984. Flexibility, Learning and Irreversibility in *Environmental Decisions, Journal of Environmental Economics and Management, 11:* 161–172.

Minns C.K., Moore J.E., Schindler D.W., and Jones M.L. 1990. Assessing the potential extent of damage to inland lakes in eastern Canada due to acidic deposition: predicting the response of potential species richness, *Canadian Journal of Fisheries and Aquatic Science 47:* 821–830.

Mitsch W.J. and Gosselink J.G. 1986. *Wetlands.* Van Nostrand Reinhold, New York.

Mitsch W.J. and Jörgensen S.E. 1989. *Ecological Engineering: An Introduction to Ecotechnology.* John Wiley and Sons, New York.

Myers N. 1988. Tropical forests and their species: going, going . . . , in Wilson E.O. ed. *Biodiversity,* National Academy Press, Washington DC: 28–37.

Myrdal G. 1975. *Against the Stream,* Vintage Books, New York.

Naiman R.J. 1988. Animal influences on ecosystem dynamics, *BioScience 38:* 750–752.

Nash C. and Bowers J. 1988. Alternative approaches to the valuation of environmental resources. In Turner R.K. ed. *Sustainable Environmental Management: Principles and Practice,* Belhaven Press, London, 118–142.

Newell R. 1988. Ecological changes in Chesapeake Bay: are they the result of overharvesting the American Oyster, *Crassostrea virginica?* In *Understanding the Estuary: Advances in Chesapeake Bay Research. Proceedings of a Conference. Chesapeake Research Consortium Publication 129,* Baltimore, Maryland.

Nichols D.S. 1983. Capacity of natural wetlands to remove nutrients from wastewater, *Journal Water Pollution Control Federation 55:* 495–505.

Nicolis G. and Prigogine I. 1977. *Self-organization in Non-equilibrium Systems: From Dissipative Structures to Order Trough Fluctuations,* John Wiley-Interscience, New York.

Nixon S.W. 1988. Physical energy inputs and the comparative ecology of lake and marine ecosystems. *Limnology and Oceanography 33:* 702–724.

Noy-Meir I. and Seligman N.G. 1979. Management of semiarid ecosystems in Israel. In Walker B.H. ed. *Management of semi-arid ecosystems,* Elsevier, Amsterdam: 113–160.

Noy-Meir I. and Walker B.H. 1986. Stability and resilience in rangelands, in Joss P.J., Lynch P.W. and Williams O.B. eds. *Rangelands: a resource under siege,* Proceedings of the Second International Rangelands Congress. International Rangelands Congress, Adelaide: 21–25.

Odum H.T. 1967. Energetics of world food production. In *The world food problem: a report of the president's science advisory committee, vol 3.* The White House, Washington, DC: 55–94.

Odum E.P. 1971. *Fundamentals of Ecology,* Saunders, Philadelphia.

Odum E.P. 1985. Trends expected in stressed ecosystems, *BioScience 35:* 419–422.

Odum E.P. 1989. *Ecology and Our Endangered Life-Support Systems.* Sinuaer Associates, Sunderland, Massachusetts.

Odum H.T. 1971. *Environment, Power and Society.* John Wiley, New York.

O'Neill R.V., DeAngelis D.L., Waide J.B., and Allen T.F.H. 1986. *A Hierarchical Concept of Ecosystems.* Princeton University Press, Princeton.

Orians G.H. and Kunin W.E. 1990. Ecological Uniqueness and Loss of Species. In Orians G.H. *et al.,* eds. *The Preservation and Valuation of Biological Resources,* University of Washington Press, Seattle, 146–184.

Paine R.T. 1966. Food web complexity and species diversity, *The American Naturalist 100:* 65–75.

Paine R.T. 1980. Food webs: linkage interaction strength and community infrastructure, *Journal of Animal Ecology 49:* 667–685.

Panayotou T. 1992. Environmental Kuznets Curve: empirical tests and policy implications, Harvard Institute for International Development, Mimeo.

Pasinetti L.L., and Solow R.M., eds. 1994. *Economic Growth and the Structure of Long-Term Development.* St. Martin's Press, London.

Pearce D.W. 1987. *Economic Values and the Natural Environment: The 1987 Denman Lecture,* Cambridge, Granta Publications.

Pearce D.W. 1988. The sustainable use of natural resources in developing countries. In Turner R.K. ed. *Sustainable Environmental Management: Principles and Practice,* Bellhaven press, London: 103–117.

Pearce D.W. 1990. An economic approach to saving the tropical forests, *LEEC Discussion Paper 90–106,* London Environmental Economics Centre, London.

Pearce D.W., Barbier E.B., and Markandya A. 1988. Environmental economics and decision-making in sub-Saharan Africa, *LEEC Paper 88–101.* London Environmental Economics Centre, London.

Pearce D.W., Barbier E.B., and Markandya A. 1990. *Sustainable Development: Economics and Environment in the Third World,* Earthscan, London.

Pearce D.W., Markandya A., and Barbier E.B. 1989. *Blueprint for a Green Economy.* Earthscan, London.

Pearce D.W. and Turner R.K. 1991. *Economics of Natural Resources and the Environment.* Harvester-Wheatsheaf, London.

Perez-Garcia J. and Lippke B.R. 1993. The timber trade and tropical forests: modelling the impacts of supply constraints, trade constraints and trade liberalisation. *LEEC Discussion Paper DP, 93–03.* London Environmental Economics Centre, London.

Perrings C. 1989. An optimal path to extinction? Poverty and resource degradation in the open agrarian economy, *Journal of Development Economics 30:* 1–24.

Perrings C. 1991. Ecological sustainability and environmental control, *Structural Change and Economic Dynamics 2:* 275–295.

Perrings C. 1992. Pastoral strategies in Sub-Saharan Africa: the economic and ecological sustainability of dryland range management, *Environment Working Paper No 57,* Environment Department, World Bank.

Perrings C. 1993. Stress, shock and the sustainability of optimal resource utilisation in a stochastic environment, in Barbier E. ed. *Economics and Ecology: New Frontiers and Sustainable Development,* Chapman and Hall, London.

Perrings C., Folke C., and Mäler K.G. 1992. The ecology and economics of biodiversity loss: the research agenda, *Ambio 30:* 201–111.

Perrings C., Mäler K-G., Folke C., Holling C.S., and Jansson B-O., eds. 1994. *Biodiversity Conservation Problems and Policies.* Kluwer Academic Publishers, Dordrecht.

Pickup G. 1991. Event frequency and landscape stability on the floodplain systems of arid Central Australia, *Quarternary Science Reviews 10:* 463–473.

Pielou E.C. 1975. *Ecological Diversity,* John Wiley, New York.

Pimm S.L. 1984. The complexity and stability of ecosystems, *Nature 307:* 321–326.

Pindyck R. 1991. Irreversibility, uncertainty and investment, *Journal of Economic Literature 29:* 1110–1148.

Poore D., Burgess P., Palmer J., Rietbergen S., and Synnott T. 1989. *No Timber Without Trees: Sustainability in the Tropical Forest,* Earthscan Publications Ltd, London.

Possingham H. and Roughgarden J. 1990. Spatial population dynamics of a marine organism with a complex life cycle, *Ecology 71:* 973–985.

Powell T.M. 1989. Physical and biological scales of variability in lakes, estuaries and the coastal ocean. In Roughgarden J., May R.M. and Levin S.A. eds. *Perspectives in Ecological Theory,* Princeton University Press, Princeton.

Puu T. 1989. *Non-Linear Economic Dynamics,* Springer-Verlag, Berlin, 1989.

Race M.S. 1986. Critique of present wetlands mitigation policies in the United States based on an analysis of past restoration projects in San Francisco Bay, *Environmental Management 1:* 71–82.

Randall A. 1991. Total and nonuse values. In Braden J.B. and Kolstad C.D. eds. *Measuring the Demand for Environmental Quality,* North Holland, Amsterdam: 303–322.

Randers J. and Meadows D. 1973. The carrying capacity of our global environment: a look at the ethical alternatives. In Daly H.E. ed. *Toward a Steady State Economy*, W.H. Freeman, San Francisco.

Rauscher M. 1990. Can cartelisation solve the problem of tropical deforestation, *Weltwirtschaftliches Archives: 126* 380–387.

Reid W.V. and Miller K.R. 1989. *Keeping Options Alive: The Scientific Basis for Conserving Biodiversity*, World Resources Institute, Washington, D.C.

Repetto R. 1986. *World Enough and Time*, Yale University Press, New Haven.

Repetto R. 1989. Economic incentives for sustainable production. In Schramme G. and Warford J.J. eds. *Environmental Management and Economic Development*, Johns Hopkins for the World Bank, Baltimore: 69–86.

Repetto R. 1990. Deforestation in the Tropics, *Scientific American, 262:* 36–45.

Repetto R. and Gillis M. 1988. *Public Policies and the Misuse of Forest Resources*, Cambridge University Press, Cambridge.

Rosser J.B. 1990. Approaches to the analysis of the morphogenesis of regional systems, *Occasional Paper Series on Socio-Spatial Dynamics 1:* 75–102.

Rosser J.B. 1991. *From Catastrophe to Chaos, A General Theory of Economic Discontinuities*, Dordrecht, Kluwer.

Roughgarden J. 1971. Density-dependent natural selection, *Ecology 52:* 453–468.

Roughgarden J. 1979. *Theory of Population Genetics and Evolutionary Ecology: An Introduction*, Macmillan, New York.

Roughgarden J. and Iwasa Y. 1986. Dynamics of a metapopulation with space-limited subpopulations, *Theoretical Population Biology 29:* 235–261.

Roughgarden J., Iwasa Y., and Baxter C. 1985. Demographic theory for an open marine population with space-limited recruitment, *Ecology 66:* 54–67.

Roughgarden J., Gaines S., and Possingham H. 1988. Recruitment dynamics in complex life cycles. *Science 241:* 1460–1466.

Roughgarden J., Pennington T., and Alexander S. 1994. Community Processes at Sea and on Land, *Proceedings of the Royal Society, Series B.* Royal Society, London.

Roughgarden J., Pennington T., Stoner D., Alexander S., and Miller K. 1991. Collisions of upwelling fronts with the intertidal zone: the cause of recruitment pulses in barnacle populations in central California. *Acta Oecologia 12:* 1–17.

Ryther J.H. 1969. Photosynthesis and fish production in the sea, *Science 166:* 72–76.

Schaeffer D.J., Herricks E.E., and Kerster H.W. 1988. Ecosystem health: measuring ecosystem health, *Environmental Management 12:* 445–455.

Schindler D.W. 1977. Evolution of phosphorus limitation in lakes: natural mechanisms compensate for deficiencies of nitrogen and carbon in eutrophied lakes. *Science 195:* 260–262.

Schindler D.W. 1980. Experimental acidification of a whole lake: a test of the oligotrophication hypothesis. In Drablos D. and Tollan E. eds. *Ecological Impact of Acid Precipitation.: Proceedings of an International Conference*, Sandefjord, Norway, March 11–14, 1980, SNSF Project, Oslo, 370–374.

Schindler D.W. 1986. The significance of in-lake production of alkalinity. *Water Air Soil Pollution 30:* 931–944.

Schindler D.W. 1987. Detecting ecosystem responses to anthropogenic stress, *Canadian Journal of Fisheries and Aquatic Science 44:* 6–25.

Schindler D.W. 1988. Experimental studies of chemical stressors on whole lake ecosystems. *Baldi Lecture. Verh. Internat. Verein. Limnol. 23:* 11–41.

Schindler D.W. 1990a. Experimental perturbations of whole lakes as tests of hypotheses concerning ecosystem structure and function. *Proceedings of 1987 Crafoord Symposium, Oikos 57:* 25–41.

Schindler D.W. 1990b. Natural and anthropogenically imposed limitations to biotic richness in freshwaters. In Woodwell G. ed. *The Earth in Transition: Patterns and Processes of Biotic Impoverishment.* Cambridge University Press, Cambridge: 425–462.

Schindler D.W. 1994. Linking species and communities to ecosystem management. *Proceedings of the 5th Cary Conference, May 1993.,*

Schindler D.W., Beaty K.G., Fee E.J., Cruikshank D.R., DeBruyn E.D., Findlay D.L., Linsey G.A., Shearer J.A., Stainton M.P., and Turner M.A. 1990. Effects of climatic warming on lakes of the central boreal forest, *Science 250:* 967–970.

Schindler D.W., Frost T.M., Mills K.H., Chang P.S.S., Davis I.J., Findlay F.L., Malley D.F., Shearer J.A., Turner M.A., Garrison P.J., Watras C.J., Webster K., Gunn J.M., Brezonik P.L., and Swenson W.A. 1991. Freshwater acidification, reversibility and recovery: comparisons of experimental and atmospherically-acidified lakes. In Last F.T. and Watling R. eds. *Acidic Deposition: Its Nature and Impacts. Proceedings of the Royal Society of Edinburgh, Vol. 97B:* 193–226.

Schindler D.W., Newbury R.W., Beaty K.G., and Campbell P. 1976. Natural water and chemical budgets for a small Precambrian lake basin in central Canada, *Journal of Fisheries Research Board Canada 33:* 2526–2543.

Schindler D.W., Newbury R.W., Beaty K.G., Prokopowich J., Ruszczynski T., and Dalton J.A. 1980. Effects of a windstorm and forest fire on chemical losses from forested watersheds and on the quality of receiving streams, *Canadian Journal of Fisheries and Aquatic Science 37 :* 328–334.

Scholes R.J. and Walker B.H. 1993. *An African Savanna. Synthesis of the Nylsvely Study.* Cambridge University Press, Cambridge (in press).

Schulze E. D. and Ulbrecht B. 1991. Acid rain – a large-scale, unwanted experiment in forest ecosystems. In Mooney H.A., Medina E., Schindler D.W., Schulze E.D. and Walker B.H. eds. *Ecosystem Experiments, SCOPE 45.* J. Wiley & Sons, New York: 89–106.

Scoones I. 1990. Why are there so many animals? Cattle population dynamics in the communal areas of Zimbabwe. Paper prepared for the Technical Meeting on Savannah Development and Pasture Production, Commonwealth Secretariat and the Overseas Development Institute, Woburn, November.

Sedjo R.A. 1992. Preserving Biodiversity as a Resource, *Resources 106:* 26–29.

Seitzinger S.P. 1988. Denitrification in freshwater and coastal marine ecosystems, *Limnology and Oceanography 33:* 702–724.

Shapley L.S. and Shubik M. 1969. On the Core of an Economic System with Externalities, *American Economic Review 59:* 678–684.

Silva, J. 1987. Responses of savannas to stress and disturbance: species dynamics. In Walker B.H. ed. *Determinants of Tropical Savannas.* IRL Press, Oxford: 141–156.

Silvander U. 1991. *The willingness to pay for angling and ground water in Sweden.* The Swedish University of Agricultural Sciences, Department of Economics.

Simon J. 1992. *Population and Development in Poor Countries,* Princeton University Press, Princeton.

Sinclair A. R. E., Olsen P.D., and Redhead T.D. 1990. Can predators regulate small mammal populations? Evidence from house mouse outbreaks in Australia, *Oikos 59* : 382–392.

Slobodkin L.B. and Sanders H.L. 1969. On the contribution of environmental predictability to species diversity. In *Diversity and Stability in Ecological Systems. Brookhaven symposium in Biology No. 22.* NTIS, Springfield, Virginia, 82–95.

Smith V.K. 1991. Household production functions and environmental benefit estimation, in Braden J.B. and Kolstad C.D. eds. *Measuring The Demand for Environmental Quality,* North Holland, Amsterdam: 41–76.

Solow R.M. 1974. Intergenerational equity and exhaustible resources, *Review of Economic Studies Symposium 41:* 29–46.

Solow R.M. 1986. On the Intergenerational Allocation of Natural Resources, *Scandinavian Journal of Economics 88:* 141–14

Solow A., Polasky S., and Broadus J. 1993. On the Measurement of Biological Diversity, *Journal of Environmental Economics and Management 24:* 60–68.

Soulé M.E. and Wilcox B.A. 1980. *Conservation Biology: an Evolutionary-Ecological Perspective.* Blackwell, Oxford.

Southwood T.R.E. 1977. Habitat, the template for ecological strategies? *Journal of Animal Ecology 46:* 337–365.

Spiller G. 1978. Hydrological nitrogen analyses. *Ecological Bulletins 28:* 107–128.

Stafford-Smith M. and Morton S. 1990. A framework for the ecology of arid Australia, *Journal of Arid Environments 18:* 255–278.

Stafford-Smith D.M. and Pickup G. 1990. Pattern and production in arid lands. *Proceedings of the Ecological Society of Australia,* 16: 195–200.

Starrett D.A. 1972. Fundamental non-convexities in the theory of externalities, *Journal of Economic Theory 4:* 180–199.

Starrett D.A. 1991. *The population externality,* Stanford, Stanford University, Department of Economics, mimeo.

Statistical Office of the United Nations 1992. *SNA Draft Handbook on Integrated Environmental and Economic Accounting, Provisional version,* Statistical Office of the United Nations, New York.

Steele J.H. 1985. A comparison of terrestrial and marine systems, *Nature 313:* 355–358.

Steele J.H. 1991. Marine functional diversity, *BioScience 41:* 470–474.

Steele J.S. 1989. A view from the oceans, *Oceanus 32:* 4–9.

Swanson T. 1992a. Economics of a Biodiversity Convention, *AMBIO* 21(3): 250–257.

Swanson T. 1992b. The Global Conversion Process, *CSERGE Discussion Paper 92–40,* CSERGE: University College London and University of East Anglia.

Swanson T. 1992c. Animal Welfare and Economics: The Case of the Live Bird Trade, in Edwards S. and Thomsen J. eds. *Conservation and Management of Wild Birds in Trade, Report to the Conference of the Parties to CITES,* Kyoto, Japan.

Swanson T. 1993a. Regulating Endangered Species, *Economic Policy,* Spring.

Swanson T. 1993b. *The International Regulation of Extinction,* Macmillan, London.

Swanson T. 1994. The economics of extinction revisited and revised. *Oxford Economic Papers, 46:* 800–821.

Swanson T. (ed.) 1995a. *The Economics and Ecology of Biodiversity's Decline,* Cambridge University Press: Cambridge.

Swanson T. (ed.) 1995b. *Intellectual Property Rights and Biodiversity Conservation,* Cambridge University Press: Cambridge.

Tongway D.J. and Ludwig J.A. 1989. Vegetation and soil patterning in semi-arid mulga lands of Eastern Australia, *Australian Journal of Ecology 15:* 23–34.

Turner R.K. 1988a. Wetland Conservation: Economics and Ethics. In Collard D. *et al.* eds. *Economics, Growth and Sustainable Environments,* Macmillan, London.

Turner R.K. ed. 1988b. *Sustainable Environmental Management: Principles and Practice,* Belhaven Press, London.

Turner R.K. 1991. Economics and Wetland Management, *Ambio 20 :* 59–63.

Turner R.K. 1992. Speculations on Weak and Strong Sustainability, *CSERGE GEC Working Paper 92–26,* CSERGE, UEA, Norwich and UCL, London.

Turner R.K., Doktor P., and Adger N. 1994. Sea level rise and coastal wetlands in the U.K.: Mitigation strategies for sustainable management. In Jansson A.M., Hammer M., Folke C. and Costanza R. (eds.) *Investing in natural capital: The ecological economics approach to sustainability.* ISEE Press/Island Press, Washington: 266–290.

Turner R.K., Kelly M., and Kay R. 1990. *Cities at Risk,* BNA International, London.

Turner R.K. and Jones T. eds. 1991. *Wetlands, Market and Intervention Failures,* Earthscan, London.

United Nations Conference on Environment and Development. 1993. Agenda 21: Programme of Action for Sustainable Development, UN, New York.

United Nations Environment Programme 1992. *The State of the Environment (1972–1992),* Nairobi.

United States Environmental Protection Agency 1982. *Chesapeake Bay, An introduction to an ecosystem,* USEPA, Washington, D.C.

United States Environmental Protection Agency (USEPA) 1983. *Chesapeake Bay, A Profile of Environmental Change,* Chesapeake Bay Program, Annapolis, Maryland.

Vane-Wright R.I., Humphries C.J., and Williams P.H. 1991. What to Protect? Systematics and the Agony of Choice, *Biological Conservation 55 :* 235–254.

Vincent J.R. 1990. Don't Boycott Tropical Timber, *Journal of Forestry 88:*56.

Vitousek P.M. 1990. Biological invasions and ecosystem processes: towards an integration of population biology and ecosystem studies, *Oikos 57 :* 7–13.

Vitousek P.M., Ehrlich P., Ehrlich A., and Matson P. 1986. Human Appropriation of the Products of Photosynthesis, *Bioscience 36:* 368–373.

Vitousek P.M. and Matson P.A. 1984. Mechanisms of nitrogen retention in forest ecosystems: a field experiment, *Science 225 :* 51–52.

Vitousek P.M. and Walker L.R. 1989. Biological invasion by Myrica faya in Hawaii, plant demography, nitrogen fixation, ecosystem effects. *Ecological Monographs 59 :* 247–265.

Volterra V. 1931. *Leçon sur la théorie mathématique de la lutte pour la vie.* Ganthier-Villars, Paris.

Wade R. 1987. The Management of Common Property Resources: Finding a Cooperative Solution, *World Bank Research Observer 2:* 219–235.

Waide J.B. and Webster J.R. 1976. Engineering systems analysis: applicability to ecosystems. In Patten, B.C. ed. *Systems Analysis and Simulation in Ecology IV,* Academic Press, New York: 329–371.

Walker B.H. 1981. Is succession a viable concept in African savanna ecosystems?

In West D.C., Shugart H.H. and Botkin D.B. eds. *Forest Succession: Concepts and Application.* Springer-Verlag, New York.

Walker B.H. 1988. Autecology, synecology, climate and livestock as agents of rangelands dynamics, *Australian Range Journal 10* : 69–75

Walker B.H., Ludwig D., Holling C.S., and Peterman R.M. 1969. Stability of semiarid savanna grazing systems, *Ecology 69* : 473–498.

Walker B.H., Matthews D.A., and Dye P.J. 1986. Management of grazing systems – existing versus an event oriented approach, *South African Journal of Science 82:* 172.

Walker B.H. and Noy-Meir I. 1982. Aspects of the stability and resilience of savanna ecosystems. In Huntley B.J. and Walker B.H. eds. *Ecology of Tropical Savannas,* Springer, Berlin: 577–590.

Walters C.J. 1986. *Adaptive Management of Renewable Resources.* McGraw Hill, New York.

Warford J.J. 1989. Environmental management and economic policy in developing countries, in Schramme G. and Warford J.J. eds. *Environmental Management and Economic Development,* Johns Hopkins for World Bank, Baltimore: 7–22.

Webb S.D. 1984. Ten million years of mammal extinctions in North America. In Martin P.S. and Klein R.G. eds. *Quaternary Extinctions,* University of Arizona Press, Tucson: 189–210.

Weisbrod B. 1964. Collective consumption services of individual consumption goods, *Quarterly Journal of Economics 77:* 189–210.

Weitzman M.L. 1992. On Diversity, *Quarterly Journal of Economics 107:* 363–406.

Weitzman M.L. 1993. What to Preserve: an Application of Diversity Theory to Crane Conservation, *Quarterly Journal of Economics, 108* : 157–184.

West D.C., Shugart H.H., and Botkin D.B. 1981. *Forest Succession: Concepts and Application.* Springer-Verlag, New York.

Westman W.E. 1985. *Ecology, Impact Assessment and Environmental Planning,* John Wiley and Sons, New York.

Westoby M., Walker B.H., and Noy-Meir I. 1989. Opportunistic management for rangelands not at equilibrium, *Journal of Rangeland Management 42:* 266–274.

Wiens 1992.

Williams M. ed. 1990. *Wetlands: A Threatened Landscape.* Basil Blackwell, Oxford.

Wills R.T. 1993. The ecological impact of Phylophthora cinnamomi in the Stirling Range National Park, Western Australia, *Australian Journal of Geology 18:* 145–160.

Wilson E.O. ed. 1988. *Biodiversity,* National Academy Press, Washington.

World Bank. 1992. *World Development Report 1992,* Oxford University Press, Oxford.

World Commission on Environment and Development 1987. *Our Common Future,* Oxford University Press, Oxford.

World Resources Institute (WRI) 1992. *World Resources 1992–93,* Oxford University Press, London.

Wright H.E., Jr. 1987. Synthesis: the land south of the ice sheets. In Ruddiman W.F. and Wright H.E. Jr. eds. *North America and Adjacent Oceans During the Last Glaciation.* DNAG Volume K-3. Geological Society of America. Boulder, Colorado: 479–488.

Index